RESIDUE REVIEWS

VOLUME 14

RESIDUE REVIEWS

Residues of Pesticides and other
Foreign Chemicals in Foods and Feeds

RÜCKSTANDS-BERICHTE

Rückstände von Pesticiden und anderen
Fremdstoffen in Nahrungs- und Futtermitteln

Edited by

FRANCIS A. GUNTHER

Riverside, California

VOLUME 14

SPRINGER SCIENCE+BUSINESS MEDIA, LLC

ISBN 978-1-4615-8412-4   ISBN 978-1-4615-8410-0 (eBook)
DOI 10.1007/978-1-4615-8410-0

© Springer Science+Business Media New York 1966

Originally published by Springer-Verlag New York Inc. 1966.

Softcover reprint of the hardcover 1st edition 1966

Library of Congress Catalog Card Number 62—18595.

Title No. 6616

Preface

That residues of pesticide and other "foreign" chemicals in foodstuffs are of concern to everyone everywhere is amply attested by the reception accorded previous volumes of "Residue Reviews" and by the gratifying enthusiasm, sincerity, and efforts shown by all the individuals from whom manuscripts have been solicited. Despite much propaganda to the contrary, there can never be any serious question that pest-control chemicals and food-additive chemicals are essential to adequate food production, manufacture, marketing, and storage, yet without continuing surveillance and intelligent control some of those that persist in our foodstuffs could at times conceivably endanger the public health. Ensuring safety-in-use of these many chemicals is a dynamic challenge, for established ones are continually being displaced by newly developed ones more acceptable to food technologists, pharmacologists, toxicologists, and changing pest-control requirements in progressive food-producing economies.

These matters are also of genuine concern to increasing numbers of governmental agencies and legislative bodies around the world, for some of these chemicals have resulted in a few mishaps from improper use. Adequate safety-in-use evaluations of any of these chemicals persisting into our foodstuffs are not simple matters, and they incorporate the considered judgments of many individuals highly trained in a variety of complex biological, chemical, food technological, medical, pharmacological, and toxicological disciplines.

It is hoped that "Residue Reviews" will continue to serve as an integrating factor both in focusing attention upon those many residue matters requiring further attention and in collating for variously trained readers present knowledge in specific important areas of residue and related endeavors; no other single publication attempts to serve these broad purposes. The contents of this and previous volumes of "Residue Reviews" illustrate these objectives. Since manuscripts are published in the order in which they are received in final form, it may seem that some important aspects of residue analytical chemistry, biochemistry, human and animal medicine, legislation, pharmacology, physiology, regulation, and toxicology are being neglected; to the contrary, these apparent omissions are recognized, and some pertinent manuscripts are in preparation. However, the field is so large and the interests in it are so varied that the editor and the Advisory Board earnestly solicit suggestions of topics and authors to help make this international book-series even more useful and informative.

"Residue Reviews" attempts to provide concise, critical reviews of timely advances, philosophy, and significant areas of accomplished or needed endeavor in the total field of residues of these chemicals in foods, in feeds, and in transformed food products. These reviews are either general or specific, but properly they may lie in the domains of analytical chemistry and its methodology, biochemistry, human and animal medicine, legislation, pharmacology, physiology, regulation, and toxicology; certain affairs in the realm of food technology concerned specifically with pesticide and other food-additive problems are also appropriate subject matter. The justification for the preparation of any review for this book-series is that it deals with some aspect of the many real problems arising from the presence of residues of "foreign" chemicals in foodstuffs. Thus, manuscripts may encompass those matters, in any country, which are involved in allowing pesticide and other plant-protecting chemicals to be used safely in producing, storing, and shipping crops. Added plant or animal pest-control chemicals or their metabolites that may persist into meat and other edible animal products (milk and milk products, eggs, etc.) are also residues and are within this scope. The so-called food additives (substances deliberately added to foods for flavor, odor, appearance, etc., as well as those inadvertently added during manufacture, packaging, distribution, storage, etc.) are also considered suitable review material.

Manuscripts are normally contributed by invitation, and may be in English, French, or German. Preliminary communication with the editor is necessary before volunteered reviews are submitted in manuscript form.

Department of Entomology   F.A.G.
University of California
Riverside, California
June 28, 1966

Table of Contents

Purification of solvents for pesticide residue analysis
By W. W. THORNBURG 1

Automated pesticide residue analysis and screening
By Professor F. A. GUNTHER and D. E. OTT 12

Biochemical and metabolic changes in plants induced by
chlorophenoxy herbicides
By D. PENNER and Dr. F. M. ASHTON 39

Review of the symposium on foreign materials in food,
Lucerne, April 8 to 9, 1965
By Dr. S. DORMAL-VAN DEN BRUEL 114

Subject Index 127

Purification of solvents for pesticide residue analysis

By

W. W. THORNBURG*

Contents

I. Introduction 1
II. Methods of purification 2
III. Commercially available solvents 3
IV. Chemical treatment of solvents 4
V. Purification of solvents by adsorption chromatography 8
VI. Conclusion 8
Summary 9
Résumé 9
Zusammenfassung 10
References 10

I. Introduction

The initial step in the analysis of pesticide residues, regardless of the method to be used, is usually the separation of the pesticide from plant or animal tissue or other organic material by solvent extraction. These extracts are usually purified by solvent partition, or elution of the pesticide from a chromatographic column with a solvent mixture. Before the final quantitation of the pesticide is made a considerable volume of solvents is usually represented. In many extraction and purification systems the impurities in the solvents, as well as the pesticides, have been concentrated.

Shortly after the introduction of the organic phosphate pesticide, parathion (0,0-diethyl-0-p-nitrophenyl phosphorothioate), AVERELL and NORRIS (1948) developed a residue method for the compound in plant tissue. The parathion was separated from plant tissue by extraction with benzene and the nitro group was reduced, diazotized, and coupled to form a magenta dye. EDWARDS (1949) found that benzene from several manufacturers contained varying amounts of impurities which gave a magenta color identical with that given by parathion. The impurity was mainly a compound with a nitro group. As the complexity and sensitivity of residue methods increases,

---

* California Packing Corporation, Emeryville, California.

the impurities introduced by chemicals used in the procedure becomes a continuing and increasingly important problem for the residue chemist.

The preferred method for the determination of most common pesticides is gas chromatography (REYNOLDS 1964). The two most common gas chromatographic procedures are the microcoulometric gas chromatograph (COULSON et al. 1960) and gas chromatography using the electron-affinity detector (LOVELOCK and LIPSKY 1960). For microcoulometric gas chromatography the solvents must be free from chlorine-and sulphur-containing compounds which can produce peaks in the pesticide range. For electron-affinity detectors the solvents must be free of any compounds which capture electrons, or affect the sensitivity of the detector. Compounds with high boiling points which have no affinity for electrons can condense on the foil and decrease the net sensitivity.

The purpose of this paper is to review methods of purifying and storing solvents for pesticide residue analysis.

II. Methods of purification

Nearly all methods of solvent purification are based on distillation, passage of the solvent thru adsorbents such as Florisil or alumina, or treatment with reactive chemicals such as sodium metal or sulfuric acid. Frequently a combination of these methods is used for in-laboratory purification.

It is to be emphasized that success of any solvent purification system depends upon the initial purity of a solvent being purified. The literature amply illustrates that one analyst may successfully purify his solvents by simple distillation, while another analyst must use an elaborate chemical treatment prior to distillation. In reviewing the literature one finds that frequently no indication is given as to the source or grade of the solvent being purified. Thus a purification system devised for a solvent by one author may not be satisfactory when applied to another batch of solvent.

The other pitfall of in-laboratory solvent purification is the inadvertent introduction of unwanted impurities into the purified solvent. WITT (1965) found that some laboratory detergents used for washing glassware introduced materials into solvents which interfered with electron-affinity analysis of pesticides. These materials were not readily removed from glassware by a water rinse.

Adsorptive materials such as Florisil or alumina are often shipped in polyethylene bags in fiber drums. The plasticizers in the bags can be adsorbed and eluted when these adsorbents are used to purify solvents. THORNBURG (1965) found that deionized water contained a series of hexane-soluble materials with electron affinity, and a response in the pesticide range.

The alternative to in-laboratory purification is the purchase of commercially purified solvents "for pesticide residue analysis."

III. Commercially available solvents

With the growing interest in the analysis of pesticides by gas chromatography several manufacturers of fine chemicals have prepared solvents for the residue analyst. Two leaders in this field at the time of writing are the Burdick & Jackson Laboratories, Inc., of Muskegon, Michigan, and the Mallinckrodt Chemical Works of St. Louis, Missouri. Generally these solvents are prepared by careful selection of high-purity industrial solvents which are then distilled in all-glass systems through efficient columns operated at high reflux-ratios. Shipping containers are one-gallon glass jugs with "Teflon" lined caps.

Petroleum and chemical companies produce several grades of each solvent to meet the needs of industry. These industrial solvents are available in 55-gallon drums, or in tank car lots. They have a high degree of purity, a closely controlled boiling range, and a low level of trace contaminants such as sulphur, aromatics, chlorides, olefins, carbonyl compounds, etc., all of which are held to a minimum. From these industrial solvents are chosen the solvents to be purified for use in pesticide residue analysis. Usually careful selection of the starting solvent negates the need for chemical pre-treatment prior to distillation.

The use of the "for pesticide residue analysis" solvents is highly recommended to the pesticide residue analyst. The problems and hazards encountered in purifying solvents are eliminated, and a source of pure solvents is assured. At the present time one manufacturer is supplying 24 distilled-in-glass solvents, many of which are useful for pesticide residue analysis.

The test of some of these solvents for "suitability for pesticide residue analysis" is performed with a gas chromatograph fitted with electron-affinity detectors. The chromatogram covers the range from the solvent peak to beyond the position of the methoxychlor peak using a standard silicone-coated column.

KNEIP (1965) indicates that one large manufacturer uses the following procedure for checking the purity of their solvents. One liter of the solvent sample is evaporated directly to about five-ml. in a Kuderna-Danish flask equipped with a three-ball Snyder column for reflux. All except petroleum ether are evaporated with the aid of an aspirator vacuum. A 100-μl. sample is then injected into a gas chromatograph equipped with an electron-affinity detector. Samples are taken several times a day to check the purity of the solvents being prepared.

A standard in a five-ml. volume is similarly sampled and tested several times during the day. The evaporation step is checked at least once a week by fortifying a liter of solvent with a standard, evaporating the liter to about five-ml., and running a 100-μl. sample.

In general, the manufacturer wishes to keep the level of impurities to less than the reading given by one to two nanograms (ng.) of heptachlor epoxide/l. of solvent. The following data were furnished by one supplier of purified solvents:

Hexane: Typical crude hexane contains by electron-affinity analysis several peaks of 50 to 70 ng./l. plus two peaks of about 200 ng./l. of contamination. After commercial purification no measurable peaks were found.

Acetone: A. R. acetone contains by electron affinity several peaks in the 10 to 20 ng./l. range with one peak of about 30 ng./l. of contamination. After commercial purification no measurable peaks were found.

Benzene: Benzene contains by electron-affinity analysis several peaks, the largest of which represents about 435 ng./l. of contamination. There were no measurable peaks after commercial purification.

Table I lists the specifications and typical analyses of several solvents specially prepared for pesticide residue analysis. These data are courtesy of the *Mallinckrodt Chemical Works.*

IV. Chemical treatment of solvents

Distillation frequently has been found to be an effective method of purification of some solvents prior to use in pesticide analysis.

NORRIS *et al.* (1954), for their colorimetric analysis of malathion [O,O-dimethyl-S-(1,2-dicarbethoxy ethyl)-phosphorodithicate], found that distillation removed an unidentified impurity from commercial carbon tetrachloride.

O'DONNEL *et al.* (1954) purified hexane for use in their colorimetric analysis of aldrin (1,2,3,4,10,10-hexachloro-1,4,4a,5,8,8a-hexahydro-1,4-*endo, exo*-5,8-dimethanonaphthalene) by simple flash distillation, discarding a ten percent forecut and leaving 15 percent bottoms. This simple distillation removed non-volatile halogen compounds or materials giving absorbance in the photometric method. These authors note that contact of the solvent with rubber should be avoided as the solvent dissolves materials which interfere. However, for many solvents, simple distillation does not give adequate purification. For example, ENOS and FREAR (1962) purified the acetonitrile used in their semi-quantitative paper chromatographic estimation of dimethoate [O,O-dimethyl S-(N-methyl carbamoylmethyl) phosphorodithioate] by distillation over concentrated sulphuric acid.

For non-reactive solvents such as hexane, pentane, and heptane, treatment with reactive chemicals followed by washing and distillation has been found to be very effective. In their colorimetric analysis of toxaphene (octachlorocamphene) GRAUPNER and DUNN (1960) purified *n*-hexane by acid treating with fuming sulphuric acid supported on dry Celite 545. The mixture was stirred for an hour, filtered through glass wool, washed with cold water, dried over anhydrous sodium sulphate, and redistilled. DUNN (1958) purified acetone and ethanol for the colorimetric analysis of Delnav [2,3-*p*-dioxanedithiol S,S-bis(O,O-diethyl phosphorodithioate)] by refluxing these solvents with 2,4-dinitrophenylhydrazine and sulphuric acid followed by distillation.

Table I. Illustrative specifications[a] for purified solvents for use in pesticide residue analysis

Test	Acetonitrile Spec. limit	Acetonitrile Typical analysis	Benzene Spec. limit	Benzene Typical analysis	Hexane Spec. limit	Hexane Typical analysis	Petroleum ether Spec. limit	Petroleum ether Typical analysis
Non-volatile material	10 p.p.m. max.	2 p.p.m.	5 p.p.m. max	1 p.p.m.	5 p.p.m. max.	1 p.p.m.	5 p.p.m. max.	0.6 p.p.m.
Total phosphorus	—	<0.001 p.p.m.	—	<0.001 p.p.m.	—	<0.001 p.p.m.	—	<0.001 p.p.m.
Density at 25°C.	0.775-0.780	0.777	—	—	0.687 max.	0.671	—	—
Water	0.3% (w/v) max.	0.04%	0.2% max.	0.005%	—	—	—	—
Acidity (as CH_3COOH)	0.03% max.	<0.003%	—	—	To pass test[b]	—	To pass test	—
Alkalinity (as NH_3)	0.0006% max.	0.0001%	—	—	—	—	—	—
Boiling range 1 ml. to 96 ml.	1.0°C. max.	0.4°C.	0.5°C. max.	0.3°C.	—	—	—	—
95 ml. to dryness	1.0°C. max.	0.1°C.	0.5°C. max.	0.3°C.	—	—	—	—
Boiling point(s)	—	—	—	—	65°-70°C.; 4.0°C. max. range	66.9°-68.4°C.; 1.5°C. range	First drop >30°C.; dry point <60°C.	37°C. 57°C.
Appearance, odor, and color	Colorless, characteristic odor	APHA <5	Clear, colorless characteristic odor	APHA<5	Colorless, faint petroleum odor	APHA <5	—	—
Thiophene	—	—	To pass test	—	To pass test	—	—	—
Other	—	—	Freezing pt. 5.2°C. min. Sulphur compounds (as S) 0.005% max. Substances darkened by sulphuric acid to pass test	5.4°C. <0.005%	Sulphur compounds (as S) 0.005%	<0.005%	Heavy oils and fats —To pass test	—

[a] No interfering G.L.C. peak greater than given by 10 ng. of heptachlor epoxide/l. of solvent. Electron-attachment detectors are used. Typical peaks observed are equivalent to 1 to 2 ng./l. (parts per trillion) as compared to heptachlor epoxide, or 15 to 30 ng./l. as compared to parathion. Total phosphorus by chemical analysis is typically less than one part per billion.
[b] Approximately 0.002 percent.

JONES and RIDDICK (1952), in their residue procedures for Dilan [2-nitro-1,1-bis(*p*-chlorophenyl)propane and butane], partitioned a hexane extract of plant tissue with acetonitrile to separate the Dilan from plant fats and waxes. Subsequently, acetonitrile has been used extensively for partition distribution cleanup of hexane extracts of plant or animal tissue.

Acetonitrile generally contains a variety of impurities and must be purified before use. The generally accepted method for the purification of acetonitrile is that described by WEISSBERGER (1955). It involves successive refluxing with phosphorus pentoxide and sodium carbonate followed by careful distillation through a good rectifying column. Acetonitrile of high purity was prepared with the use of this method by JANZ and DANYLUK (1959) who refluxed the acetonitrile with phosphorus pentoxide six times for 36 hours each time before continuing with the sodium carbonate treatment and distillation. O'DONNEL *et al.* (1965) prepared very high purity acetonitrile by a different procedure. They distilled acetonitrile over anhydrous sodium carbonate and potassium permanganate. The distillate was made slightly acid with concentrated sulphuric acid and distilled through a 40-plate column.

Many chlorinated pesticides, when heated with diphenylamine in the presence of anhydrous zinc chloride, form a blue or green colored complex. CUETO (1960) used this procedure for the analysis of dieldrin (1,2,3,4,10, 10-hexachloro-6,7-epoxy - 1,4,4a,5,6,7,8,8a - octahydro - 1,4-*endo,exo*-5,8-dimethanonaphthalene) in animal fat. Ethyl ether was used as the solvent for the zinc chloride. Peroxide-free ether was prepared by treating the solvent with ferrous sulphate or sodium bisulphite solutions. The ether was washed with water, dried with sodium sulphate, and redistilled.

Reaction with sodium metal followed by distillation has been favored as a purification procedure for aliphatic hydrocarbons. Chlorine-containing compounds are removed and the solvent is dried by this procedure. BOWERY *et al.* (1959), for the analysis of insecticides in tobacco, purified *n*-pentane by distilling commercial grade *n*-pentane over sodium metal in an all-glass apparatus, discarding a five percent forecut and the 15 percent bottoms.

SCHECHTER *et al.* (1965) found that dispersed sodium, because of the tremendous surface exposed, is more effective and rapid for removing impurities from some solvents than sodium wire, ribbon, or shot. These authors refluxed hydrocarbon solvents with a small amount of 50 percent dispersed sodium in mineral oil; the solvent was then distilled.

Distillation with sodium can also be used to purify solvents other than hydrocarbons. BANN *et al.* (1958), for their analysis of endrin (*endo,endo* isomer of dieldrin), purified isopropanol and petroleum ether by distillation over sodium metal in an all-glass apparatus.

It should be emphasized that care should be used in handling sodium metal. After distillation is complete the sodium can be destroyed by the cautious addition of isopropyl alcohol. Dispersed sodium is safe when covered by a solvent or inert gas, but it may ignite if exposed to air, especially if the solvent is removed.

Some analysts have found that treatment with sodium hydroxide will purify solvents adequately. BOWAN and BEROZA (1965 a), for their determination of extraction "p"-values refluxed hexane, heptane, and iso-octane over sodium hydroxide and distilled before use. These authors (1965 b) used the same procedure for the purification of hexane for the colorimetric and electron-affinity analysis of Imidan (O,O-dimethyl S-phthalimidomethyl phosphorodithioate).

For some procedures it is necessary to remove aldehydes and ketones from solvents. BLINN et al. (1954) prepared aldehyde- and ketone-free methanol by refluxing the solvent with 2,4-dinitrophenylhydrazine and a small amount of hydrochloric acid. BEROZA (1956) in his procedure for the determination of methylenedioxyphenyl synergists used in fly sprays prepared aldehyde-free methanol by refluxing it with 5,5-dimethyl-1,3-cyclohexanedione.

AKERMAN et al. (1963), in their spectrophotometric analysis of PCNB (pentachloronitrobenzene), purified petroleum ether by prewashing with three portions of concentrated sulphuric acid. The petroleum ether was then washed with distilled water, then with two percent sodium bicarbonate solution, and finally with distilled water until such washes were neutral to pH indicator paper. The petroleum ether was dried with anhydrous sodium sulphate and filtered. The filtrate was distilled through a 400-mm. column of two-mm. glass beads. The initial 100-ml. forecut and 100-ml. tail product from a four-l. batch were discarded. The distillate was stored in glass bottles. This product must then meet the specification: "Upon concentration of 800 ml. of reagent grade petroleum ether to 5 ml., sufficient interference shall not be obtained to produce a reagent blank equivalent to more than 3.6 ± 0.8 µg. apparent PCNB. Nor shall such a concentrate, to which a 5-ml. aliquot of a standard purified petroleum ether solution containing 50 µg. of PCNB was added, and subsequently concentrated to 5 ml., deviate more than $\pm$ 10% in absorbence compared value to a similar standard not containing the concentrate of 800 ml. of petroleum ether."

ORDAS et al. (1956), in their colorimetric analysis of heptachlor (1,4,5, 6,7,8,8-heptachloro-3a,4,5,5a-tetrahydro-4,7-endo-methanoindene) and chlordane (2,3,4,5,6,7,8,8-octachloro-2,3,3a,4,7,7a-hexahydro-4,7-methanoindene), purified pentane by washing it with concentrated sulphuric acid, followed by treatment with 1.0 N potassium hydroxide solution saturated with potassium permanganate. The solvent was finally distilled over potassium hydroxide pellets. This purified solvent must pass the following interference test: "Two liters of solvent, evaporated in a 50°C. water bath through a three-bulb Snyder column to near dryness and made to react with Polen-Silverman or Davidow reagent as described in their procedure, shall not exhibit color in excess of the equivalent of 1.5 µg. of heptachlor or chlordan, nor shall there be any apparent residue from 2 liters of pentane when it is taken to dryness at 50°C."

V. Purification of solvents by adsorption chromatography

For some analytical procedures solvents can be purified by percolation through a column of adsorptive material. Adsorptives used are usually silica gel, alumina, or Florisil. For example, BRUCE *et al.* (1958) purified petroleum ether and *n*-heptane by allowing the solvents to percolate through a column of silica gel. These solvents were then used for the residue analysis of diphenylamine.

BLINN *et al.* (1954), in a procedure for chlorobenzilate (ethyl *p,p'*-dichlorobenzilate), removed aromatics from light petroleum ether by passage through a column of silica gel. ROSENTHAL *et al.* (1957) in their method for the fungicide Karathane [2-(1-methyl-*n*-heptyl)-4,6-dinitrophenyl-crotonate] purified chloroform, hexane, and acetonitrile by passage through alumina. They note that when opening cans containing solvent that the lip of the can should be cleaned well with solvent to remove lacquer, soldering flux, and the like before use. This advice should be given consideration when handling any solvent to be used for any type of pesticide residue analysis.

GORDON *et al.* (1962), in their method for the analysis of perthane [1,1-dichloro-2,2-bis(*p*-ethylphenyl)ethane], purified both *n*-hexane and acetonitrile. The technical grade hexane was purified by passage through activated alumina. These authors found that with a column 1.5 inches in diameter one pound of activated alumina would clean up two gallons of solvent. Reagent grade acetonitrile was purified in the same manner. Technical grade acetonitrile was distilled prior to passage through the column. GORDON *et al.* (1961) used a similar procedure to purify dimethyl formamide used in the colorimetric analysis of Lethane 384 (2-butoxy-2'-thiocyanodiethyl ether). ROSENTHAL *et al.* (1957) also used a similar procedure for purification of *n*-hexane, acetonitrile, and chloroform used in the analysis of the fungicide dinitrocaprylphenylcrotonate.

VI. Conclusion

The purity of an in-laboratory purified solvent depends upon the purity of the original solvent, treatment, and care which is taken during purification. Exposure to rubber, plastic, or detergent-contaminated glassware can negate hours of effort taken in the cleanup of a solvent.

The use of reactive chemicals such as metallic sodium always incurs some hazard to the laboratory worker. Taking all factors into consideration, the purchase of commercially purified solvents made expressly for pesticide residue analysis is strongly recommended.

As the market for these solvents expands the number of solvents available should increase, and the amount of impurities remaining in the solvents should decrease. These solvents are somewhat expensive, and if a large volume of solvent is to be used in the laboratory for an extended period of time, several sources and grades of solvents can be investigated. For many analyses several solvents will often be equally satisfactory. For example, methylene chloride can be substituted for the more expensive and toxic chloroform, or for carbon tetrachloride.

Summary

The first step in pesticide analysis is usually extraction of the sample with a solvent, and for most analytical techniques the solvent must be purified prior to use. Nearly all methods of solvent purification are based on distillation, passage of the solvent thru adsorbtive mixtures, treatment of the solvent with reactive chemicals, or a combination of these techniques. The alternative to in-laboratory purification is the purchase of commercially purified solvents prepared especially for use in pesticide residue analysis.

Care must be taken with any solvent to avoid contact with rubber, plastics, or any material which can cause contamination that would interfere with pesticide residue analysis. The avoidance of contamination is especially important in gas chromatographic techniques using microcoulometric or electron-affinity detectors.

The purchase of commercially purified solvents has many advantages over in-laboratory purification. The use of such solvents is especially recommended if it is anticipated that a relatively small amount of solvent will be needed for an analytical procedure.

Résumé *

La première opération dans l'analyse des pesticides est généralement l'extraction de l'échantillon au moyen d'un solvant et pour la pluoart des méthodes analytiques le solvant doit être au préalable purifié. Presque toutes les méthodes de purification des solvants sont basées sur la distillation, le passage au travers de mélanges adsorbants, le traitement des solvants par des réactifs chimiques ou une combinaison de ces techniques. A la purification au laboratoire il n'est d'autre alternative que l'achat de solvants purs du commerce spécialement préparés pour les analyses des résidus de pesticides.

Avec tout solvant il convient d'éviter le contact avec le caoutchouc, les matières plastiques ou tout autre matière pouvant introduire une souillure capable d'interférer avec l'analyse des résidus de pesticides. La prévention des contaminations est particulièrement importante en analyse par chromatographie gazeuse avec les détecteurs microcoulométriques et à capture d'électrons.

L'achat de solvants purs du commerce présente de nombreux avantages sur la purification au laboratoire. L'emploi de tels solvants est particulièrement recommandé lorsque la technique à suivre ne nécessite qu'une quantité relativement faible de solvant.

---

* Traduit par R. MESTRES.

Zusammenfassung*

Der erste Schritt in der Analyse von Pestiziden ist gewöhnlich die Extraktion des Musters mittels eines Lösungsmittels, das bei den meisten analytischen Methoden vor Gebrauch gereinigt werden muss. Praktisch alle Methoden zur Reinigung des Lösungsmittels basieren auf Destillation, Durchlauf des Lösungsmittels durch adsorptive Gemische, Behandlung des Lösungsmittels mit reaktiven Chemikalien oder auf der Kombination dieser Techniken. Als Alternative zu einer Reinigung im Labor können kommerziell gereinigte Lösungsmittel gekauft werden, die speziell für die Rückstandsanalyse von Pestiziden hergestellt werden.

Es muss streng darauf geachtet werden, dass Lösungsmittel nicht mit Gummi, Plastik oder irgendeinem verunreinigenden Material in Berührung kommen, die Pestizid-Rückstandsanalysen stören könnten. Es ist besonders wichtig, Verunreinigungen zu vermeiden, wenn Methoden angewandt werden, die mit microcoulometrischen oder elektronischen Affinitätsdetektoren arbeiten.

Der Ankauf von im Handel befindlichen gereinigten Lösungsmitteln hat viele Vorteile gegenüber der Reinigung im Labor. Die Verwendung von solchen Lösungsmitteln ist besonders zu empfehlen, wenn vorauszusehen ist, dass nur eine relativ kleine Menge Lösungsmittel für eine analytische Methode gebraucht wird.

References

AKERMANN, H. J., L. J. CARBONE, and E. J. KUCHAR: Modifications to the spectrophotometric analysis of PCNB (Terraclor) in soil and crops. J. Agr. Food Chem. 11, 297 (1963).

AVERELL, P. R., and M. V. NORRIS: Estimation of small amounts of O,O-diethyl O-p-nitrophenyl thiophosphate. Anal. Chem. 20, 753 (1948).

BANN, J. M., S. C. LAU, J. C. POTTER, J. W. JOHNSON, Jr., A. E. O'DONNELL, and F. T. WEISS: Determination of endrin in agricultural products and animal tissues. J. Agr. Food Chem. 6, 196 (1958).

BEROZA, M.: Determination of methylenedioxyphenyl-containing synergists used in analysis of fly sprays. J. Agr. Food Chem. 4, 53 (1956).

BLINN, R. C., F. A. GUNTHER, and M. J. KOLBEZEN: Microdetermination of the acaricide ethyl p,p'-dichlorobenzilate (Chlorobenzilate). J. Agr. Food Chem. 2, 1080 (1954).

BOWMAN, M. C., and M. BEROZA: Analysis of Imidan colorimetrically and by electron-affinity gas chromatograph. J. Assoc. Official Agr. Chemists 48, 922 (1965 b).

— — Extraction p-values of pesticides and related compounds in six binary solvent systems. J. Assoc. Official Agr. Chemists 48, 943 (1965 a).

BOWERY, T. G., W. R. EVANS, F. E. GUTHRIE, and R. L. RABB: Insecticide residues on tobacco. J. Agri. Food Chem. 7, 693 (1959).

BRUCE, R. B., J. W. HOWARD, and J. B. ZINK: Determination of diphenylamine residues on apples. J. Agr. Food Chem. 6, 597 (1958).

COULSON, D., L. A. CAVANAGH, J. E. DE VRIES, and B. WALTHER: Microcoulmetric gas chromatography of pesticides. J. Agr. Food Chem. 8, 399 (1960).

* Übersetzt von H. MARTIN.

CUETO, C., JR.: Colorimetric determination of dieldrin and its application to animal fat. J. Agr. Food Chem. 8, 273 (1960).

DUNN, C. L.: Determination of 2,3-p-dioxanedithiol S,S-bis(O,O-diethyl phosphorodithioate). J. Agr. Food Chem. 6, 203 (1958).

EDWARDS, F. I., JR.: Sources of error in estimating small amounts of parathion. Anal. Chem. 21, 1415 (1949).

ENOS, H. F., and D. E. H. FREAR: Method for the detection of microgram quantities of O,O-dimethyl-S-(N-methylcarbamoylmethyl) phosphorodithioate (dimethoate) in milk. J. Agr. Food Chem. 10, 477 (1962).

GIANG, P. A.: Fluorometric method for estimation of residues of Bayer 22408. J. Agr. Food Chem. 9, 42 (1961).

GORDON, C. F., L. D. HAINES, and I. ROSENTHAL: Analytical method for determining 1,1-dichloro-2,2-bis(p-ethylphenyl) ethane in rat fat and cow's milk. J. Agr. Food Chem. 10, 380 (1962).

— —, and A. L. WOLFE: A procedure for the microdetermination of 1-butoxy-2-(2-thiocyanoethoxy)-ethane (Lethane 384) with applications for determination of residues in milk and animal tissues. J. Agr. Food Chem. 9, 478 (1961).

GRAUPNER, A. J., and C. L. DUNN: Determination of toxaphene by a spectrophotometric diphenylamine procedure. J. Agr. Food Chem. 8, 286 (1960).

JANZ, J. J., and S. S. DANYLUK: Hydrogen halides in acetonitrile. I. Ionization processes. J. Amer. Chem. Soc. 81, 3846 (1959).

JONES, L. R., and J. A. RIDDICK: Separation of organic insecticides from plant and animal tissues. Anal. Chem. 24, 569 (1952).

KNEIP, T. J.: Personal communication (1965).

LOVELOCK, J. E., and S. R. LIPSKY: Electron affinity spectroscopy—a new method for the identification of functional groups in chemical compounds separated by gas chromatography. J. Amer. Chem. Soc. 82, 431 (1960).

NORRIS, M. V., W. A. VAIL, and P. R. AVERELL: Colorimetric estimation of malathion residues. J. Agr. Food Chem. 2, 570 (1954).

O'DONNELL, A. E., H. W. JOHNSON, JR., and F. T. WEISS: Chemical determination of dieldrin in crop materials. J. Agr. Food Chem. 3, 757 (1955).

—, M. M. NEAL, F. T. WEISS, J. M. BANN, T. J. DE CINO, and S. C. LAU: Chemical determination of aldrin in crop materials. J. Agr. Food Chem. 2, 573 (1954).

O'DONNELL, J. F., J. T. AYERS, and C. K. MANN: Preparation of high purity acetonitrile. Anal. Chem. 9, 1611 (1965).

ORDAS, E. P., V. C. SMITH, and C. F. MEYER: Spectrophotometric determination of heptachlor and technical chlordan on food and forage crops. J. Agr. Food Chem. 4, 444 (1956).

REYNOLDS, H. L.: Use of gas chromatography by Food and Drug Administration for pesticide residue analysis. J. Gas Chromatog. 2, 219 (1964).

ROSENTHAL, I., C. F. GORDON, E. L. STANLEY, and M. H. PERLMAN: Microdetermination of the fungicide dinitrocaprylphenylcrotonate in food crops and animal tissues. J. Agr. Food Chem. 5, 914 (1957).

SCHECHTER, M. S., C. CORLEY, and J. E. FAHEY: Purification of solvents for pesticide residue analysis and electron-capture gas chromatography. Unpublished (1965).

THORNBURG, W. W.: Preparation and extraction of samples prior to pesticide residue analysis. J. Assoc. Official Agr. Chemists 48, 1023 (1965).

WEISSBERGER, A., E. PROSKAUER, J. RIDDING, E. TOOPS, eds.: Technique of organic chemistry. Vol. VII, 2nd ed. New York: Interscience 1955.

WITT, J. M.: Personal communication (1965).

Automated pesticide residue analysis and screening*

By

F. A. GUNTHER** and D. E. OTT**

Contents

I. Domestic regulation of residues 12
II. International regulation of residues 14
III. Analytical screening for residues 15
IV. Automated screening for residues 18
V. General residue-analytical requirements 19
VI. Automated residue-analytical requirements 20
VII. AutoAnalyzer*** applications to residue assay 23
    a) Organophosphorus compounds and cholinesterase inhibitors . . . 23
    b) Biphenyl 26
Summary 34
Résumé 34
Zusammenfassung 35
References 36

I. Domestic regulation of residues

In the United States, with probably the most intense and diversified agriculture in the world, there are registered for permitted use more than 900 chemical compounds[1] in more than 60,000 pesticidal formulations (FREEMAN 1965). In this country uses of these materials on more than 2,500 crop items (KIRK 1964) and other foodstuffs are regulated by joint considerations of the *U. S. Food and Drug Administration* (F.D.A.) and the *U.S. Department of Agriculture* (U.S.D.A.) (HARRIS and CUMMINGS 1964), with assignments of tolerances or permitted residue values for amounts of particular chemicals to be legally acceptable on and in particular raw agricultural commodities at time of sale or consumption. These values may range from an actual zero to more than 100 parts per million (p.p.m.), depending in large measure upon the pharmacology and toxicology of the candidate chemical. Until recently and in broadest terms "zero" as inter-

---

* Presented in part at the September 1965 Technicon symposium "Automation in Analytical Chemistry," New York, N. Y.

** Department of Entomology, University of California, Riverside 92502.

*** Technicon Controls, Inc., Ardsley (Chauncey), New York 10502.

[1] All pesticide chemicals mentioned in text are identified chemically in Table IV.

preted by regulatory agencies could mean either numerical tolerance denied, because of probable hazard to be associated with the projected use, or a number less than unity predicated upon residue analytical capabilities and credulity at the moment. For certain pesticide chemicals which were of widespread application on certain major crops or products, and which were also of pharmacological concern, as residue analytical methodology improved in minimum detectability, and in reliability, some "zero" and other tolerance values have been lowered with certain commodities (e.g., milk) to accommodate a new analytical capability; present tolerances often reflect residues that *could* remain at harvest. This completely unsatisfactory dynamic tolerance situation has existed particularly with several of those pesticides which contain organically bound halogens, such as dieldrin, endrin, and heptachlor. It has also been involved both legally and morally with numerous chemicals previously registered by the U.S.D.A., and with F.D.A. concurrence, on a "no-residue" basis (HARRIS and CUMMINGS 1964): at the time the "no-residue" registration was granted finite residues were not demonstrable yet subsequently analytical detection of persisting residues was made possible by improvements in methodology, usually through exploitation of modern analytical instrumentation, and the original specific registration was revoked.

Clearly, legally permitted amounts of pesticide residues in foods and feeds based upon analytical capabilities constitute an untenable concept, yet pertinent legislation and regulations have not permitted any other ready recourse. A recent report from the *National Academy of Sciences* (*National Academy of Sciences—National Research Council 1965*) recommends adoption of "permitted residues" for those situations presently covered by tolerance values and "negligible residues" to cover instances of inadvertent and otherwise negligible contamination (e.g., drift of spray to adjoining crops), with both concepts based upon pharmacological considerations alone, with complete exclusion of minimum detectability analytical considerations. Adoption of this basis will immediately stabilize the requirements imposed upon the residue analysts who monitor our food and feed supplies; they annually analyze hundreds of thousands of samples of hundreds of different food items for several hundred pesticide chemicals for large-scale farmers and most agricultural production organizations, for segments of both the raw and the processed food-producing industries, and for both state and federal agricultural and regulatory agencies. Also, both domestic and irrigation waters are coming within this purview as possible carriers of pesticide chemicals for direct ingestion by man and animals or for possible contamination of crops. Since many of these chemicals are persistent for maximum efficiency and economy in the practical pest control so essential today, they can contaminate agricultural soils for long periods, thus invoking analytical efforts to establish soil residues across the country in those instances where certain of these chemicals can migrate from soil and soil water into some growing crops.

II. International regulation of residues

Routine analytical and legislative interests in pesticide residues are not confined to the United States, as shown in Table I.

Table I. *Examples of legislative control of pesticide residues in foodstuffs[a]*

Country	Residue control program	Timing restrictions	Sources of residue data
Australia	State jurisdiction[b]	Optional[c]	State
Austria[d]	Federal law	Occasional[e]	State
Belgium	Regulated by decree[b]	Regulated	State, institutes
Canada[k]	Compulsory, comprehensive[b]	Regulated[e]	Applicant, State
Denmark[f]	State jurisdiction	None	National Pesticide Laboratory-Ministry of Agriculture
France	Restricted by decree[b]	Regulated[c]	State, universities
Great Britain[g]	Voluntary, new chemicals	Regulated[c]	Public analysts
Greece	Compulsory (olives, citrus [b,m])	Regulated	State, institutes[m]
India	State jurisdiction	Optional[c]	State
Indonesia	State jurisdiction	—[e]	State
Israel	State jurisdiction	Regulated	State, institutes
Italy	Compulsory (olives, grain)	Regulated[e]	Provinces
Japan[h]	None[c]	Regulated	State
Lebanon	State jurisdiction[b,c]	—[e]	Institute[m]
New Zealand	Regulated by law	Regulated	State
Norway	None[b]	Probable[c]	Institutes
Spain	None[b]	None	Institutes
Sweden	State jurisdiction[b]	Regulated[c]	State
Switzerland	Regulated[b]	Regulated[c]	Cantons
The Netherlands[i]	Compulsory, comprehenve[j]	Regulated	State institutes
Turkey	Government advisors[b]	Occasional[e]	State[m]
United Arab Republic	Comprehensive	Regulated	Ministry of Agriculture
United States[l]	Compulsory, comprehensive[b]	Regulated	Various
U.S.S.R.	Compulsory, comprehensive	Regulated	National Commission
W. Germany	Comprehensive	Probable	States, institutes, industry

[a] Largely from GUNTHER (1962) except where otherwise indicated.
[b] Certain materials prohibited.
[c] Extensive revision anticipated or in progress.
[d] BERAN (1963).
[e] Not by statute, but minimum interval is often recommended on label.
[f] BRO-RASMUSSEN (1965).
[g] MARTIN (1963) and MILLER (1965).
[h] SUZUKI (1963).
[i] *Netherlands legislation* (1965).
[j] Complete revision of old legislation.
[k] Tolerances for aldrin, dieldrin, and heptachlor revoked, as reported editorially in Food Chem. News **7** (16), 23 (July 19, 1965).
[l] Several tolerances recently revised in the organochlorine insecticide group.
[m] CARMAN (1965).

The Austrian, British, and Japanese laws and regulations pertaining to pesticide residues in foodstuffs have been discussed in detail elsewhere (BERAN 1963, MARTIN 1963, SUZUKI 1963, MILLER 1965). For several years Germany has been designing appropriate legislation with contemplated "tolerance" values purportedly often fractions of ours. The Netherlands has just enacted residue legislation (*Netherlands Legislation* 1965); 39 of their typical "tolerance" values are compared with ours and some others in Table II. Some materials prohibited in these countries are permitted in the United States, and *vice versa*.

As stressed by DORMAL and HURTIG (1962) and mentioned by GUNTHER (1962) and by WILSON and BAIER (1963), tolerance value differences among countries can cause difficulties in international trade as well as in the food industries and in the residue laboratory. Because of differences in basic foods, in total diets, and in available foods it must be conceded that complete international uniformity of tolerances is hardly possible nor even desirable, yet gross discrepancies among international tolerances can and will prevent exchanges of some commodities. For example, whichever country has the larger tolerance for a particular compound in a given foodstuff will be required routinely to re-evaluate the product analytically to assure its conformity with a lesser importing tolerance requirement of a customer nation. Possible trade barriers of this sort will require extensive and routine residue analyses by both the exporter and the importer, the former to assure initial sale and the latter to assure compliance with local regulations and thus to permit resale.

III. Analytical screening for residues

In the present usage, pesticides are those chemicals successfully and commercially used in combating the pests that interfere with the production, transport, and storage of agricultural crops and products. Although there are more than 900 of these chemical compounds, they may be loosely categorized as follows into frequently overlapping groups or types according to common elements in their largely organic composition (although of reviving importance, the familiar inorganic pesticides are excluded here):

Antimony (i.e., potassium antimony tartrate)
Arsenic[2] (e.g., the nitrophenylarsonic acids)
Bromine (e.g., methyl bromide, Dibrom)
Chlorine (e.g., DDT and many others)
Copper (e.g., the organic copper fungicides)
Mercury (e.g., the organic mercury fungicides)
Nitrogen (as nitro groups, carbamate moieties, substituted ureas, triazines, etc.)
Phosphorus (e.g., malathion, parathion)
Sulfur (e.g., parathion, Thiram)
Tin (i.e., Fentin)

[2] Prohibited in W. Germany.

Table II. *Examples of existing tolerance values for food crops in various countries*

Pesticide chemical	The Netherlands[a]	U.S.A.	Other
Aldrin	0.1	0.1-0.25	Zero (U.S.S.R.)[b] Restricted (Great Britain)[c]
Arsenates (as As)	1	2.3	1-2 (Canada)[b]
Biphenyl (citrus fruits)	30	110	70 (Great Britain)
Captan	15	100	—
Carbaryl	3	10	—
Chlorbenside	1.5	3	—
Chlordane	0.1	0.3	—
Chlorofenson (Ovex)	1.5	3	—
Copper	10 (50 on celery leaf)	Exempt	—
DDT	1	7	0-1 (U.S.S.R.)[b]
Diazinon	0.5	0.75-1	0.25-0.75 (Canada)[b]
Dicofol (Kelthane)	2	10	—
Dieldrin	0.1	0.1-0.25	Restricted (Great Britain)[c]
Dioxathion	1	2.8	—
Dodine	1	5	—
Endosulfan	0.5	2	—
Ethion	0.5	1	—
Fentin	1	—	—
Ferbam	7	7	—
Folpet	20	50	—
Heptachlor	0.1	Zero	Restricted (Great Britain)[c]
Lindane	2	10	—
Linuron	0.2	1	—
Malathion	3	8	—
Maneb	7	10	—
Mercury	0.03	Zero	—
Methoxychlor	10	14	—
Methyl parathion	0.5	1	0.75 (Switzerland)[b]
Nicotine	1	2	—
Parathion	0.5	1	1 (Canada)[b]
Pyrethrum (total)	3	1	—
Ronnel	0.4	0-0.5	—
SOPP[d] (citrus fruits)	30	10	—
Tetradifon	3	2-5	—
Thiometon	0.5	—	0.1-0.5 (Switzerland)[b]
Thiram	3	7	—
Toxaphene	0.4	1-7	—
Zineb	7	7	—
Ziram	7	7	—

[a] Netherlands legislation 1965.
[b] DORMAL and HURTIG (1962).
[c] MILLER (1965).
[d] Sodium o-phenylphenate.

Total residue methods for these elements or groups therefore constitute screening (segregative) methods for large numbers of pesticide chemicals. In addition, the organophosphorus and the carbamate pesticides all inhibit to varying degrees the nerve-impulse mediator cholinesterase (ChE), thus providing another "broad-broom" analytical handle.

Since the production of a given crop usually requires the use of more than one pesticide chemical—and some crops may require several during a particularly bad pest season—monitoring and other interested residue laboratories are perforce required to eliminate analytically all but a few of the theoretically hundreds of possibilities[3], then to prove natures and amounts above tolerances of those actually present; often, too, the plant or animal metabolic alteration products of the original pesticide chemical are more toxic than the parent compound and sometimes must also be determined in routine analyses.

It is clear, then, that routine screening procedures to demonstrate the presence or tolerance-realistic "absence" of groups or categories of pesticides, as above, are becoming increasingly important. With an analytical target value (i.e., a stable tolerance value), screening would partition a set of samples into those more than about ten percent above tolerance from those at or below tolerance. From long experience in the United States, the last category would contain more than 97 percent (KIRK 1964, BENDER 1965) of all samples representing good agricultural practice during production of the crop. The remaining few aberrant samples or sample extracts could then be re-examined with more definitive methods incorporating adequate resolution, identification, and measurement.

Considering that in a single state (California) several thousands of samples, and that in a single country (the United States) many thousands of samples, are routinely analyzed annually by the regulatory agencies[4], these screening procedures can become overwhelming burdens as the current very small percentage of commercial shipments examined is necessarily increased. As pointed out earlier, in the United States many agencies and groups are involved in routine pesticide residue analyses at the microgram level both for screening and for evaluative purposes; these agencies and groups include many branches of the federal and state governments, segments of the food industries, the manufacturers of pesticide chemicals, private laboratories, and most state agricultural experiment stations. The result has been an acute shortage of analytical personnel adequately trained in this

---

[3] A knowledge of crop production practices eases this burden by elimination of illogical possibilities for a given season and a given geographical area.

[4] For example, in 1963 F.D.A. alone analyzed 29,244 domestic and 832 import samples, with illegal pesticide residues in 2.1 percent of the former and 0.1 percent of the latter; in the first half of 1964, F.D.A. analyzed 17,123 domestic and 360 import samples, with 2.9% of the former and 1.9 percent of the latter containing illegal residues (KIRK 1964). Also, the California Department of Agriculture annually analyzes at least 12,000 samples of foods, feeds, and other animal products, with similar small percentages of illegal residues (BENDER 1965).

new and exacting area (GUNTHER 1962). With the addition of the new and intense nation-wide interest in monitoring soils, waters, the "market basket," and people for pesticide and other man-made chemical residues, this dearth of analysts will become even more acute. Since residue chemistry in its present state is completely empirical in nature (GUNTHER 1962), it becomes extremely important to circumvent in some manner this immediate costly shortage of analysts for residue screening purposes—empiricism exists upon both mistakes and successes, but analytical inadequacies in this area can noisily condemn a necessary and safe pest-control agent or set of samples as well as mistakenly and quietly reprieve a potentially harmful material or set of samples.

IV. Automated screening for residues

One obvious solution to help circumvent both these problems is to automate the routine residue analyses to the point where the available skilled analysts can devote more of their time to those incisive analytical efforts always associated with aberrant samples, and also to the ever-increasing evaluative residue programs required under current legislative restrictions imposed to maintain quality and quantity in our air, food, fiber, and water supplies. Continuous soil-residue surveillance is a necessary and ambitious part of this scheme because of persistent residues that may migrate into growing crops.

Of the approximately 400 major agricultural pesticidal chemicals available today, approximately half contain organically bound bromine (several) or chlorine (many), and about one-third are the remarkably useful (CASIDA 1964) cholinesterase-inhibiting organophosphorus compounds; the remaining types of compounds are extremely varied in nature (see Table IV) but include several nitrogen-containing compounds including a number of carbamates, the insecticidal members of which are also cholinesterase (ChE) inhibitors. Many of the thiolo- and thionophosphorus pesticides are converted to much more potent ChE inhibitors by selective oxidation of the involved sulfur atom to yield a phosphate. It is thus clear that automated screening procedures for these two halogens, for phosphorus, for sulfur, for nitrogen, and for ChE activity (OTT and GUNTHER 1966 b) before and after $P \rightarrow S$ to $P \rightarrow O$ oxidation, for example, would be immediately and immensely helpful in providing, *in toto*, definitive residue information. In the United States tolerance values range up to 50 p.p.m. (as inorganic bromide) for the organobromine compounds, and up to eight p.p.m. for the ChE inhibitors; with the exception of the bromine-containing pesticides, however, most tolerance values are in the vicinity of unity. Automated screening for chlorine, for phosphorus, for sulfur, for nitrogen, and for unoxidized-oxidized ChE activity at any level in the range 0.1 to 10 p.p.m., chosen to approximate the tolerance value of the most toxic compound being sought, would separate the samples into below-tolerance categories *versus* possible illegal categories. The former would be of no more concern whereas the latter would

1ave to be investigated—sometimes in considerable detail—for adequate
:stablishment of illegality, and the collective "numbers" for the above param-
:ters would often almost pinpoint a particular pesticide.

The inherent advantages of reliability, reproducibility, and speed of the
1utomated analytical system are essential to the successful production of ade-
¡uate numbers of routine residue analyses in these escalating international
·esidue surveillance programs in the interests of trade as well as of the public
1ealth. As mentioned earlier, it is to be expected (KIRK 1964) that less than
1bout two percent of the samples amenable to this penta-parameter examina-
:ion would probably require other, close, and definitive examination
(GUNTHER and BLINN 1955, SCHECHTER and HORNSTEIN 1957, GUNTHER
1962, ZWEIG 1963-1964) firmly to establish natures and magnitudes of
1bove-tolerance levels of particular compounds.

V. General residue-analytical requirements

As discussed in somewhat different terms elsewhere (e.g., GUNTHER 1962,
ZWEIG 1963-1964), all pesticide residue analyses of foodstuffs involve seven
basic steps:

1. *Sampling* to represent maximum load present or to represent average load present. These two objectives utilize widely different sampling techniques.
2. *Subdivision* of sample to homogeneous and representative aliquot.
3. *Processing, or* transfer of sought chemical plus coextractives into
   suitable solvent or solvent mixture. This is usually a concentration
   operation (i.e., bulk of sample vs. bulk of final solvent concentrate).
4. *Cleanup,* or isolative operations:
   a) Segregation of sought chemical or a suitable derivative of it from
      the bulk of the coextractives.
   b) Further segregation as necessary to accommodate the specificity
      and other critical requirements of the determinative method
      utilized.
5. *Determination* of sought chemical by any combination of biological[5],
   chemical, or physical procedures adequate for the purpose.
6. *Calculations* and comparisons with data from controls, fortified controls, and standards.
7. *Interpretation of data.* The major contingency here involves the availability of suitable control (untreated) samples for comparative purposes unless acceptable compensation is possible.

---

[5] Mention should be made of the use of conditioned reflexes, especially of cats, as
a biological screening technique for pesticide residues (MEDVED' *et al.* 1964).

VI. Automated residue-analytical requirements

It is to be expected that automated residue analyses would ordinarily start with step 4 above, although step 3 would be ideal and will be possible in some instances, as illustrated later with the citrus fungistat biphenyl. Automated determinative steps only would start with step 5.

Cleanup can involve many operations and stages, depending upon the amount and the chemical and physical properties of the sought chemical, the natures and amounts of the substrate extractives present, and the determinative method selected. Frequently, adequacy of the cleanup establishes which determinative method(s) can be used satisfactorily.

As has been discussed in detail elsewhere (GUNTHER and BLINN 1955, GUNTHER 1962, ZWEIG 1963-1964, FREHSE 1964), cleanup procedures can be combinations of both chemical and physical manipulations to take advantage of almost any property unique to the sought chemical in the presence of the substrate extractives; if such a property does not conveniently exist, it is often possible to convert the sought residue into an exploitable derivative or other alteration product. Automated analysis modules[6] already exist for conducting some types of column chromatography, solvent partitionings, filtrations, ion-exchange separations[7], hydrolyses, oxidations, reductions, steam and other distillations, dye-body formations, dialyses, etc.; many of these modules can be adapted to perform most of the usual types of cleanup operations required here. Residue cleanup procedures not yet completely automated would include gas[8], paper, and thin-layer chromatography, vacuum sublimations, and a few others; scraped-off "spots" from thin-layer chromatograms can be analyzed automatically, however (OTT and GUNTHER 1966 c), and the same technique should apply to paper "spots."

If cleanup is adequate, determinative methods are almost legion. The method of choice then depends upon the degree of specificity desired and the minimum detectability (SUTHERLAND 1965) required. Collectively, the available pesticide chemicals represent an immense diversity of chemical types or classes and include inorganic, metallo-organic, and organic compounds. There already exist some AutoAnalyzer[6] methods for functional groups, releasable anions and cations, and other moieties that undoubtedly could be adapted to pesticide residue analyses with only minor modifications after suitable cleanup. In Table III are listed some examples of these promising applications, with examples of the pesticide chemicals that could be examined AutoAnalytically. The organochlorine pesticides, for example, have been mentioned frequently. Not only are there many of them, but also they are of considerable pharmacological concern. Since they all contain organically bound chlorine many residue and residue screening methods determine

---

[6] Technicon Controls, Inc., Ardsley (Chauncey), New York 10502.

[7] A technique apparently first applied to pesticide residue analyses by WESTENBERG (1954) and recently reviewed in this application by CALDERBANK (1965).

[8] Barber-Coleman Company (Rockford, Illinois) has just announced "a completely automated laboratory gas chromatograph."

Table III. *Existing AutoAnalyzer[6] methods[a] probably directly adaptable to pesticide residue methods with suitable cleanup*

Analytical entity	Example of pesticide	Reaction to obtain analytical entity
Aldehydes	Kelthane	Fujiwara
Ammonia	Ammonia	None
Chloride ion	Many	Combustion; wet oxidation
Chlorine gas	Many	Wet oxidation
Copper ion	Several fungicides	Acidification
Diphenylamine[b]	Diphenylamine	None
Hydrogen cyanide	Hydrogen cyanide	None
Hydrogen sulfide	Several	Reduction; oxidation[c]
Ketones	Several	Dehydrohalogenation, oxidation
Lead ion	Several insecticides	Acidification; precipitation[c]
Nicotine[d]	Nicotine	None
Phenols	Many carbamates	Hydrolysis
Phosphate ion	Many	Hydrolysis, oxidation
Sulfur dioxide	Aramite	Hydrolysis
Tin ion	Fentin	Hydrolysis
Triazines[e]	Several herbicides	Colorimetry
Urea[f]	Several fungicides and herbicides	Hydrolysis
Zinc ion	Several fungicides	Acidification

[a] Technicon Methodologies (see footnote six in text).
[b] GERKE (1964).
[c] BLINN and GUNTHER (1961).
[d] The Technicon AutoAnalyzer Bibliography, 1964 ed.
[e] A method[a] for urea based upon its reaction with diacetyl to form triazine, which is determined colorimetrically.
[f] May work for some substituted ureas.

the releasable chloride ion or chlorine gas, after removal of inorganic chlorides in the isolates, with interpretation in terms of the most toxic organochlorine compound probably present or suspect. Both wet chemical and combustion methods have been carefully and extensively evaluated for these purposes, as reviewed by GUNTHER and BLINN (1955) and ZWEIG (1963-1964). As indicated in Table III, there are several excellent automated procedures for determining both chloride ion and chlorine gas, however obtained. A more extensive tabulation can easily be made for likely, known, analytically-used reactions of many of the common pesticide chemicals, reactions readily adapted to existing AutoAnalyzer modules and which incorporate "built-in" cleanup. For example, the SCHECHTER-HORNSTEIN residue method (see GUNTHER and BLINN 1955 for details) for benzene-hexachloride involves dechlorination of this compound to benzene which is distilled into a nitrating mixture to yield largely m-dinitrobenzene which, after extraction, is converted to a violet-red complex salt with alkali and methyl ethyl ketone.

Automated modules for determinative operations include colorimeters, electrophoresis units, visible and ultraviolet spectrophotometers, fluorometers, flame photometers, and assemblies of modules for microtitrations (e.g., metered titration reagents for a species such as chloride ion as commonly obtained by combustion) or completely automated combustion and "micro-titrations" by direct potentiometry of the combustion or otherwise released chloride ion from organochlorine pesticides (GUNTHER et al. 1965). Electrophoretic separations have been used occasionally to separate and sometimes to determine plant and animal metabolites of enzyme inhibitor pesticides (BRUAUX et al. 1964) or radiolabeled pesticides (METCALF 1965). AutoAnalyzer electrophoresis is presently an all-liquid determinative system with no provision for isolation of "zones" but may still find special and useful application here. Not yet automated are infrared spectrometry (BLINN and GUNTHER 1962-1963 and 1964) and micropolarography (GAJAN 1964), both unusually valuable tools in residue chemistry because they not only measure but also help characterize the determined species.

As pointed out earlier, the degree of cleanup achieved must be adequate for the purpose at hand. Clearly, fractional p.p.m. determinations will require (GUNTHER 1962, p. 224) scrupulous minimization of, or acceptable compensation for, the always present and always variable background (GUNTHER and BLINN 1955) contributions from substrate, solvents, reagents, and equipment, including spurious signals from recorders. Most practical determinations in the five-to-100 p.p.m. range will be less exacting in this requirement. If suitable control (untreated) samples are available, analytical comparisons among adequate replicates of control, fortified control (GUNTHER 1962), and treated samples will usually suffice providing there has been no storage deterioration of the sought compound. Control samples in monitoring programs are rarely available, however, and only an experienced analyst can interpret the resultant data unless a carefully planned statistical program of sampling, samples, and analyses has been involved. In these other instances, recourse may be had to such acceptable means of compensation as split-blank procedures, fortified versus unfortified samples in preparation of standard curves, dual analyses such as hydrolyzable p-nitrophenol content plus ChE activity for parathion residues, for example, and others (GUNTHER 1962). Screening programs for individual functional group analysis such as chlorine, bromine, nitrogen, phosphorus, or sulfur substituents, carbamate moieties, thiono- or thiolo-phosphate components, etc., can be interpreted only in terms of the most toxic compound thought to be present; if properly combined, however, these screening techniques can provide a high degree of exclusion in most applications to the more legally interesting pesticide chemicals, as pointed out earlier.

The remainder of this review is concerned with examples of applications of AutoAnalyzer analytical techniques to pesticide residue problems.

VII. AutoAnalyzer applications to residue assay

a) Organophosphorus compounds and ChE inhibitors

A promising, but as yet almost unexplored, approach to automated screening for total organophosphorus residues lies in existing AutoAnalyzer systems: the wet digestion-oxidation technique for total organic phosphorus (WINTER and FERRARI 1964) and its modification (WEINSTEIN et al. 1964). In our laboratories we are exploring this application of the further modified system shown in Figure 1, and have achieved 70-plus percent recoveries with reproducibilities of ± 5 percent in the range of two-to-ten μg./ml. of "cup" solution with a mixture of perchloric and sulfuric acids as digestant and with several insecticides[9]. For evaluation of most crops it will be necessary to do some cleanup prior to AutoAnalysis with this system. As mentioned earlier, the automated analysis of thin-layer "scrapings" has been developed (OTT and GUNTHER 1966c) and it is anticipated that thin-layer chromatography will provide, in most instances, sufficient cleanup as well as some specificity for final automatic analysis by this wet digestion technique.

A logical approach to the earlier suggested simple paired combinations of screening techniques would be to split the sample stream automatically and to analyze for total organically bound phosphorus, while simultaneously Auto-Analyzing the other portion of the sample stream for ChE inhibition properties (first suggested in OTT and GUNTHER 1966b). Thus, with a dual-pen recorder, two separate and valuable interdependent pieces of information could be recorded simultaneously from each sample.

A flow diagram depicting an existing AutoAnalyzer system for the measurement of ChE inhibition properties is shown in Figure 2. This is a slight modification (OTT and GUNTHER 1966b) of the original method of WINTER (1960), and has been employed in actual (e.g., OTT and GUNTHER 1966a) and several simulated (OTT and GUNTHER 1966b) residue problems. Typical results obtained with this system from a simulated residue problem involving market canned peaches fortified with technical grade parathion are shown in Figure 3.

A further modification to this anti-ChE system which simplifies it to a single-pump method is shown in Figure 4. Technicon Sampler I and a

---

[9] Manuscript submitted for publication.

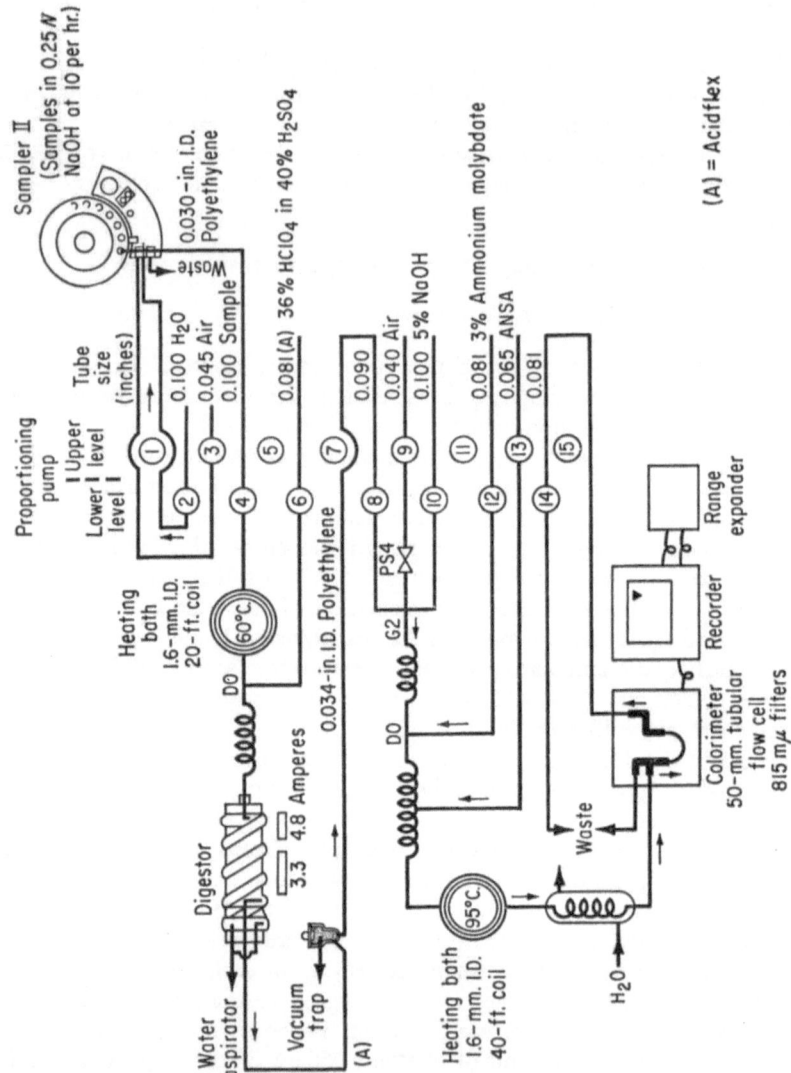

Fig. 1. Automated wet digestion system for total organically bound phosphorus

constant level vessel were used here; however, the newer Sampler II can be used effectively after modification of the sampler plate according to the method of STRANDJORD and CLAYSON (1964).

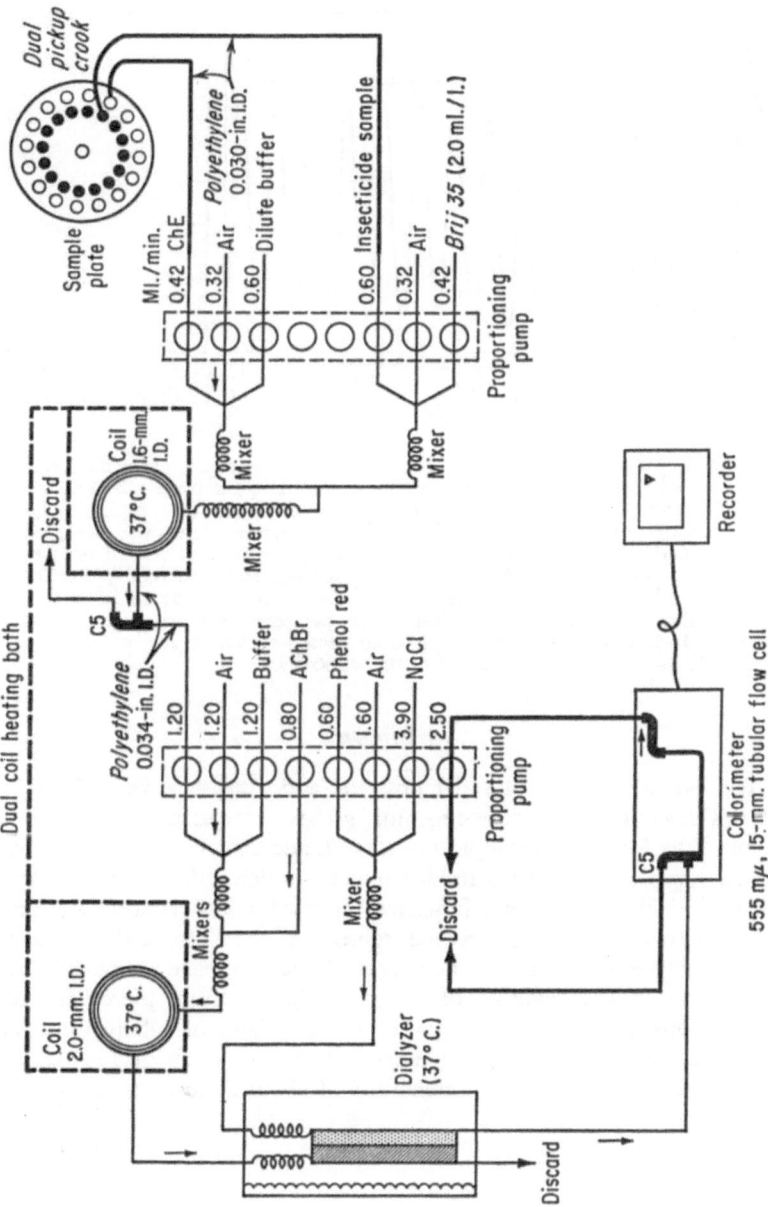

Fig. 2. Automated system for determining ChE inhibition properties. Redrawn in part from OTT and GUNTHER (1966 b). Heavy lines and italic print indicate changes from the original WINTER (1960) system

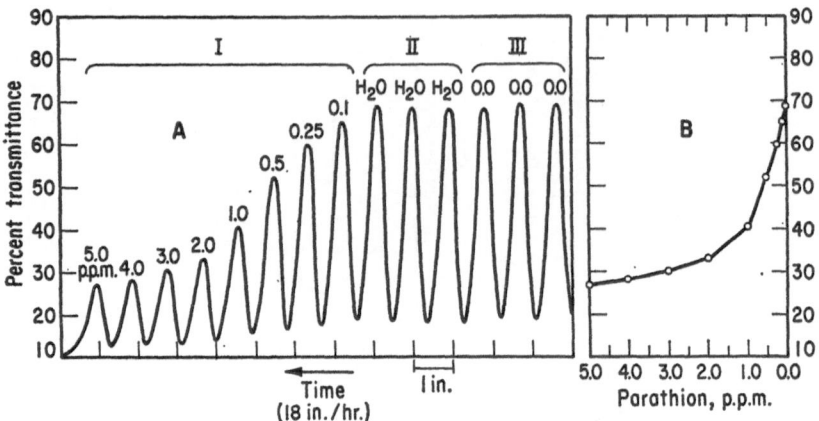

Fig. 3. *A.* Representative recording of market canned peaches, unfortified and fortified with technical grade parathion: (I) aliquots from *n*-hexane stripping
solutions of peaches fortified in terms of p.p.m., (II) replicate checks of
ChE activity (human plasma) when water alone was sampled, and
(III) aliquots from unfortified stripping solutions of canned peaches. Each
aliquot was equivalent to 25 g. of peaches and after final preparation the
analytical solution represented five g./ml. Chart speed: 18 in./hr.

B. Standard curve of technical grade parathion-fortified peaches drawn on
Technicon chart reader from chart record shown in *A.*

(Redrawn from OTT and GUNTHER 1966 b)

b) Biphenyl

The preceding examples demonstrate and suggest some of the possibilities for automated screening for groups or classes of pesticides as
residues. The following example (GUNTHER and OTT 1966) will show what
has been done to automate a residue method which will screen for one particular pesticide residue, the fungistat biphenyl which concentrates in the
rind of citrus fruits. This method represents the "ultimate" in pesticide
residue analysis for it is the first example of complete automation from
homogenization (processing) of the raw substrate from pieces of rind to
read-out between samples eight to 11 minutes later of relative amount of
residue on a recorder. The useful range is from one to about 150 p.p.m., on
a whole-fruit basis, with a reproducibility of about three percent.

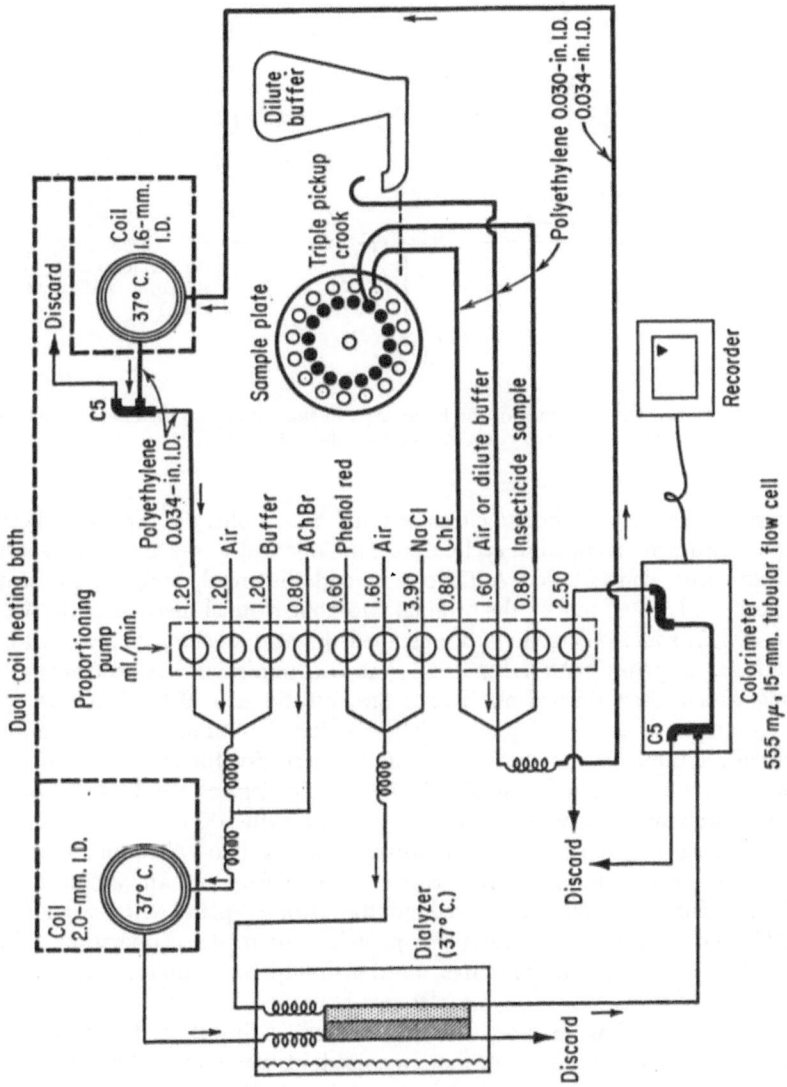

Fig. 4. Single proportioning pump automated system for ChE inhibition properties (redrawn from OTT and GUNTHER 1966 b)

A basic flow diagram of the major steps involved in this method is shown in Figure 5. The rind sample is automatically homogenized in water,

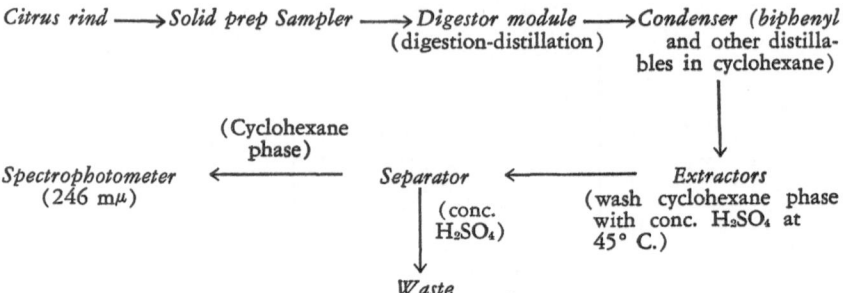

Fig. 5. Simplified flow diagram of the basic steps in the automated system for biphenyl residues in citrus fruit rind (Gunther and Ott 1966)

then steam distilled to liberate citrus oils and waxes plus the biphenyl. These steam volatiles are trapped in cyclohexane, the oils and waxes are continuously extracted into concentrated sulfuric acid and discarded, and the biphenyl is determined at 246 mμ. In the original two-hour manual method (Gunther *et al.* 1963), some *p*-cymene either escaped the acid washing or was formed from other terpenes (Gunther 1962), and to minimize was oxidized with permanganate then washed out with more sulfuric acid. With Valencia and other oranges, lemons, and grapefruit the present continuous method either essentially eliminates *p*-cymene without the necessity for an oxidation step or else the small-size starting sample effectively negates troublesome background components; most manual methods start with 100 to 150 g. of whole fruit whereas the present method utilizes one- to two-g. samples of rind.

One- or two-g. samples of manually chopped citrus rind are weighed into Solidprep Sampler cups and placed on the sampler plate with two empty "wash"[10] cups between each sample cup. With this mode of operation, using a standard factory-set programmer in the Solidprep Sampler, the average rate of sampling is six actual samples per hour.

The detailed flow diagram for the method is shown in Figure 6. The sample of rind is automatically homogenized in water to a thin puree, an aliquot of which is steam distilled in the presence of heat-concentrating sulfuric acid to destroy cellular material and thus to liberate the biphenyl plus citrus oils (mostly terpenoids) and waxes. These steam volatiles are trapped in a stream of cyclohexane, the oils and waxes are quantitatively extracted into a stream of fresh concentrated sulfuric acid, intimately mixed with the

---

[10] Needed to space the samples sufficiently to permit complete return to baseline between each peak. For peaks of strong absorbance three "wash" cups may be desired, but for routine screening of samples of nearly equivalent biphenyl content a single "wash" cup should suffice.

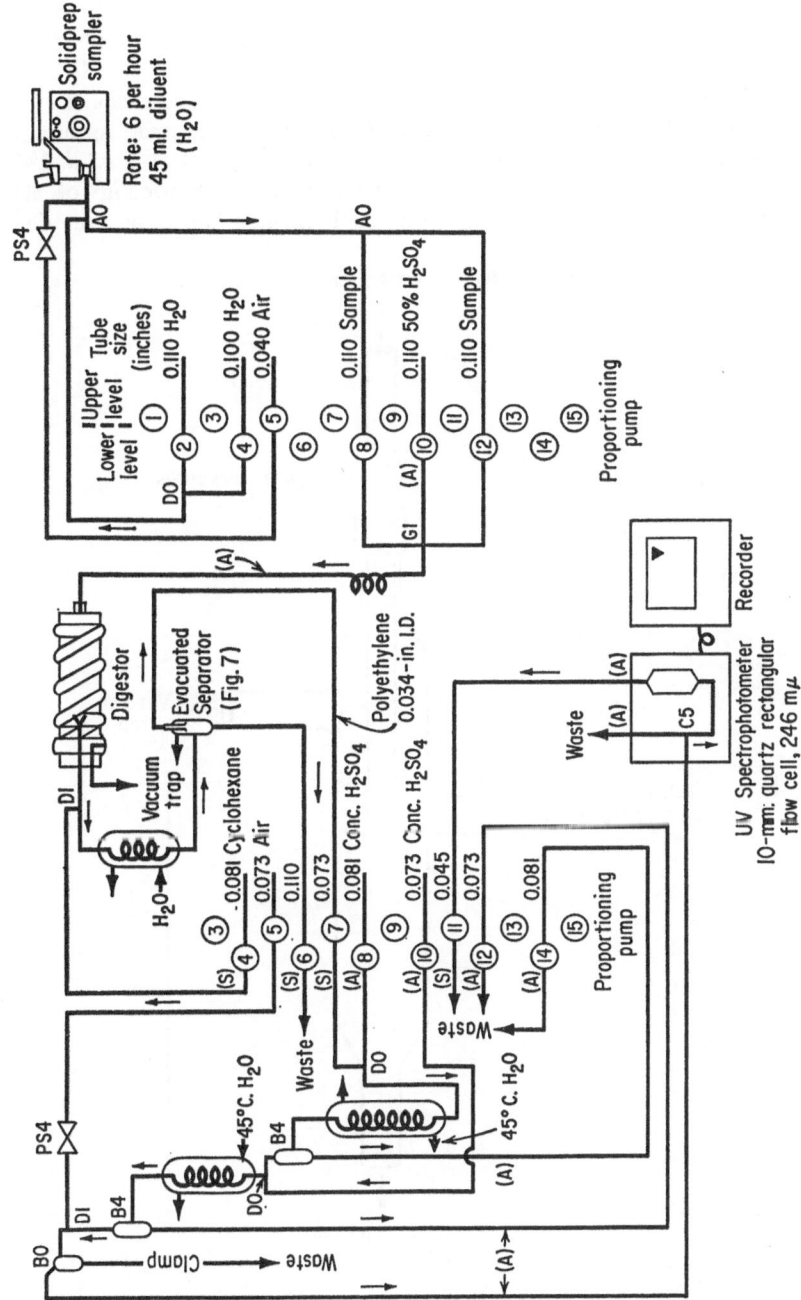

Fig. 6. Automated system for biphenyl residues in citrus fruit rind; *A*, Acidflex tubing and *S*, Solvaflex tubing (redrawn from GUNTHER and OTT 1966)

cyclohexane stream, and discarded; the biphenyl left in the cyclohexane stream is measured continuously in a recording dual-beam spectrophotometer by its strong absorption at 246 mμ.

This system utilizes either reagent grade or Spectrograde cyclohexane for extraction of the distillate and continuously fresh, reagent-grade sulfuric acid for the cyclohexane wash series; the manual method (GUNTHER *et al.* 1963) requires use of Spectrograde cyclohexane exclusively. A vacuum of about two inches of mercury is required on the Evacuated Separator (see details

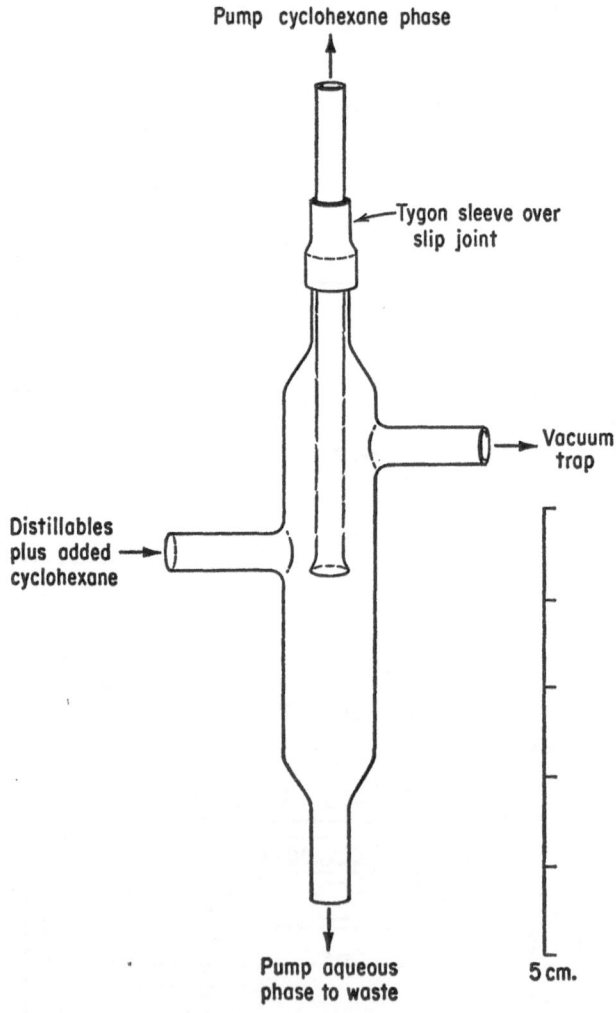

Fig. 7. Details of Evacuated Separator essential to the biphenyl-residue automated system (redrawn from GUNTHER and OTT 1966)

in Fig. 7) which in turn supplies the vacuum to collect the vapors in the collection funnel inserted into the end of the Digestor. One of the major obstacles in the development of this automated procedure was the problem of separating an aqueous phase from an immiscible solvent phase while both phases are under reduced pressure. This obstacle was overcome with the development of this special glass Evacuated Separator. Since it promises to be of broad utility in other automated systems, it is shown in detail.

Typical chart recordings are reproduced in Figure 8 as obtained from

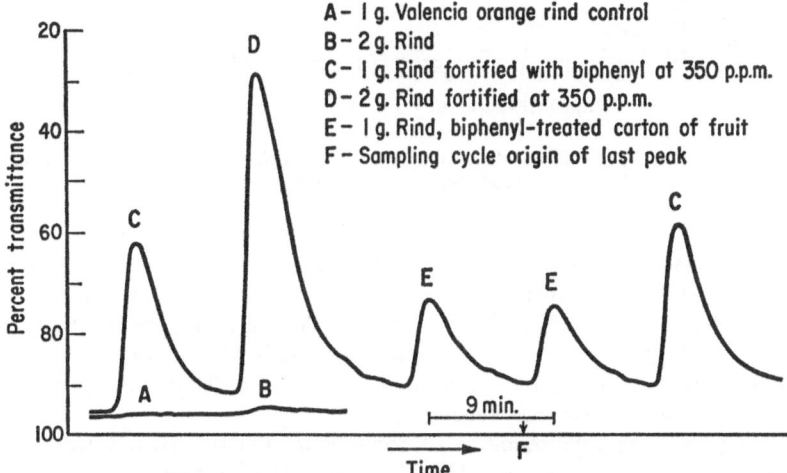

Fig. 8. Typical chart recordings obtained from the biphenyl-residue automated system (redrawn from GUNTHER and OTT 1966)

one- or two-g. starting samples of chopped Valencia orange rind. They are from the system operated for seven consecutive, noise-free hours during which 34 samples were run with no operator attention except to load cups. The average time interval for an analysis was thus about 11 minutes/test exclusive of the warm-up and shut-down operations. Attention is called both to the absence of significant background (Fig. 8) from even two g. of control rind and also to the close agreement between replicates when comparisons are made among ordinate intercepts of peak maxima alone.

Absorbance-unit peak height measurements of this type for a series of controls fortified with from 35 to 700 p.p.m. of biphenyl in 95 percent ethanol established a fortified-control standard curve (Fig. 9). Each point in this figure is the mean value from at least two replicates. The lowest point represents 35 p.p.m. in the rind, a value roughly equivalent to seven p.p.m. based on the U. S. *Food and Drug Administration* practice of weight of whole fruit[11] and well below The Netherlands' tolerance of 30 p.p.m. as

---

[11] Mature Valencia oranges contain 18.7±6.3 percent rind based upon 297 measurements.

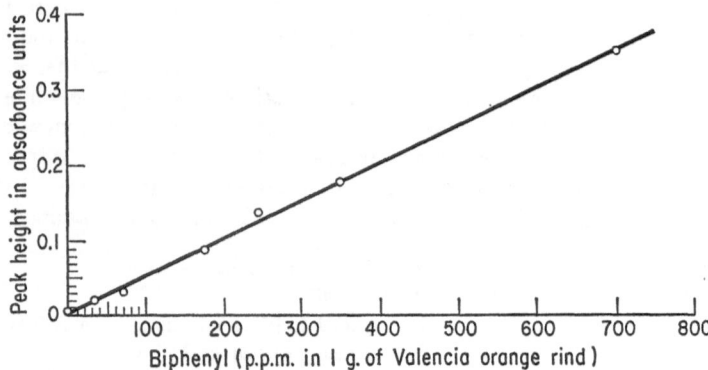

Fig. 9. Standard curve from AutoAnalyzed biphenyl-fortified samples of Valencia orange rind controls (redrawn from GUNTHER and OTT 1966)

earlier listed in Table II; replicated six times it yielded a mean value of 0.017 ± 0.001 absorbance unit. The value at 70 p.p.m., replicated four times, gave 0.037 ± 0.004 absorbance unit, and three replicates at the 175 p.p.m. level gave 0.082 ± 0.005 absorbance unit. These values were not corrected for the uniform background response of 0.003 absorbance unit from one-g. lots of control Valencia rind.

Samples more representative of commercial practice than fortified controls were also AutoAnalyzed. Half of a field box of tree-ripened Valencia oranges were stored six days at 25° to 30°C. in a standard vented citrus fruit shipping carton with three standard, commercial, biphenyl-treated liners separating layers of fruit. The other half box was kept in a separate room as control. Fruits from the "treated" carton produced the two peaks labeled E in Figure 8. From the fortified-control standard curve there were 175 ± 8 p.p.m. of biphenyl in and on the rind of this treated fruit, or 33 ± 2 p.p.m. on a whole-fruit basis. It was not possible directly and simply to relate the slope of the fortified control standard curve to that of a primary standard curve with biphenyl alone, for without a "keeper" such as orange oil biphenyl was lost in large and variable amounts during the Solidprep homogenization cycle.

With this system only minutes are required for an analysis as contrasted with the hours per sample required by the several manual methods presently being used both in the United States and in Europe (RAJZMAN 1965).

Table IV. *Precise chemical designations for pesticide chemicals mentioned in text according to accepted American or European usage*

Common or trivial name	Precise name
Aldrin	1,2,3,4,10,10-hexachloro-1,4,4a,5,8,8a-hexahydro-1,4-*endo*, *exo*-5,8-dimethanonaphthalene
Aramite	2-(*p*-tert-butylphenoxy) isopropyl-2-chloroethyl sulfite
Benzenehexachloride	1,2,3,4,5,6-hexachlorocyclohexane
Biphenyl	diphenyl (Europe)
Captan	*N*-trichloromethylmercapto-4-cyclohexane-1,2-dicarboximide
Carbaryl	1-naphthyl *N*-methylcarbamate
Chlorbenside	4-chlorobenzyl-4'-chlorophenyl sulfide
Chlordane	2,3,4,5,6,7,8,8-octachloro-2,3,3a,4,7,7a-hexahydro-4,7-methanoindene
Chlorfenson	*p*-chlorophenyl-*p*'-chlorobenzenesulfonate
DDT	1,1,1-trichloro-2,2-bis(*p*-chlorophenyl)ethane
Diazinon	*O,O*-diethyl-*O*-[2-isopropyl-4-methylpyrimidinyl-(6)]-phosphorothioate
Dibrom	1,2-dibromo-2,2-dichloroethyl dimethyl phosphate
Dicofol	1,1-bis(*p*-chlorophenyl)-2,2,2-trichloroethanol
Dieldrin	1,2,3,4,10,10-hexachloro-6,7-epoxy-1,4,4a,5,6,7,8,8a-octa-hydro-1,4-*endo,exo*-5,8-dimethanonaphthalene
Dioxathion	2,3-*p*-dioxanedithiol-*S,S*-bis(*O,O*-diethylphosphorodithioate)
Dodine	*N*-dodecyl-guanidine acetate
Endosulfan	6,7,8,9,10,10-hexachloro-1,5,5a,6,9,9a-hexahydro-6,9-methano-2,3,4-benzodioxathiepin-3-oxide
Ethion	*O,O,O',O'*-tetraethyl-*S,S'*-methylene-bis(phosphorodithioate)
Fentin	diphenyltin
Ferbam	ferric dimethyldithiocarbamate
Folpet	*N*-(trichloromethylthio) phthalimide
Heptachlor	1,4,5,6,7,8,8-heptachloro-3a,4,7,7a-tetrahydro-4,7-methanoindene
Kelthane	see Dicofol
Lindane	gamma-isomer of benzenehexachloride
Linuron	3-(3,4-dichlorophenyl)-1-methoxy-1-methylurea
Malathion	*O,O*-dimethyl-*S*-(1,2-dicarbethoxy ethyl)-phosphorodithioate
Maneb	manganese ethylenebisdithiocarbamate
Methoxychlor	1,1,1-trichloro-2,2-bis(*p*-methoxyphenyl)ethane
Methyl parathion	dimethyl analog of parathion
Nicotine	1-methyl-2-(3'-pyridyl)-pyrrolidine
Parathion	*O,O*-diethyl-*O*-*p*-nitrophenyl-phosphorothioate
Pyrethrum	pyrethrum
Ronnel	*O,O*-dimethyl-*O*-(2,4,5-trichlorophenyl) phosphorothioate
SOPP	sodium *o*-phenylphenate
Tetradifon	2,4,5,4'-tetrachlorodiphenyl sulfone
Thiometon	*O,O*-dimethyl-(2-ethylthioethyl)-phosphorodithioate
Thiram	tetramethylthiuram disulfide
Toxaphene	a chlorinated camphene containing 67 to 69 percent chlorine
Zineb	zinc analog of Maneb
Ziram	zinc dimethyldithiocarbamate

Summary

Residues of pesticides in the total environment are of legal and moral concern to individuals everywhere and to increasing numbers of governmental agencies. Tremendously expanded efforts to establish, on a continuing basis, natures and magnitudes of all types of pesticide residues in this environment—including air, clothing, foodstuffs, soil, and water—have emphasized the problems of residue analytical uncertainties at the required microgram levels and of the dearth of trained residue analysts. The former is a matter of both personnel and equipment and arises from the great variety of compounds and substrates involved as well as from the complexities of the literally hundreds of details of techniques exploitable for the particular purpose. The latter is a matter of time and the establishment around the world of training centers adequate to attract chemists into this area.

Both problems can be circumvented in large part by automation of major and basic residue analytical procedures. For example, of the more than 900 pesticidal compounds extant, at least 400 are major agricultural pesticides and among these about half contain organically bound chlorine and about a third are organophosphorus compounds which are also cholinesterase inhibitors. Some of these also contain organically bound sulfur, and many pesticides are nitrogen heterocycles or contain nitro-, amino-, or carbamate functions. Automated screening procedures for examining routine samples for these five parameters would therefore be immensely helpful in segregating possible above-tolerance samples for other, close, and definitive examination if necessary.

Two basic procedures—total organically bound phosphorus determination and cholinesterase inhibition assay—have been developed and are discussed to illustrate the flexibility and promise in this field of existing automated step methodologies, including solvent partitioning, hydrolysis, destructive oxidation, dialysis, dye-body formation, and colorimetry. A rapid and reproducible automated screening procedure for biphenyl in citrus fruit rind is also discussed in detail to illustrate possibilities for development of these procedures for specific pesticide residues; this application involves automated homogenization and extraction of rind, steam distillation, solvent partitioning, reaction-solvent partitioning, and ultraviolet spectrophotometry.

Résumé*

Les résidus de pesticides dans la nature présentent partout un problème d'intérêt légal et moral dont le public et les agences gouvernementales s'inquiètent de plus en plus. D'extrêmement grands efforts ayant pour but d'établir d'une manière continue la nature et la magnitude de toutes sortes de résidus de pesticides dans notre milieu—y compris l'air, les vêtements, les produits alimentaires, le sol et l'eau—ont accentué les problèmes des incertitudes

---

* Traduit par G. P. et C. L. GEORGHIOU.

analytiques des résidus au niveau requis de microgrammes ainsi que ceux dûs au manque de chimistes spécialisés dans l'étude des résidus. Le premier problème est une question à la fois de personnel et d'instruments et provient de la grande variété des composés et des substrats qui entrent en question tout aussi bien que des complexités des centaines de détails de technique pouvant s'employer dans un but déterminé. Quant au deuxième problème, il s'agit ici d'une question de tempts et également de l'établissement de part le monde de centres de formation adéquats afin d'attirer des chimistes dans cette branche.

Les deux problèmes pourraient se résoudre en grande partie par l'emploi d'un système automatique pour les analyses qui sont fondamentales et d'importance majeure. Par example, parmi plus de 900 composés pesticides, il y en a au moins 400 qui sont des pesticides de grande importance à l'agriculture et parmi ceux-ci, à peu près la moitié contiennent du chlore joint au carbone et environ un tiers sont des composés organo-phosphorés qui sont également des inhibiteurs de cholinesterase. Certains de ces pesticides contiennent du souffre joint au carbone et beaucoup sont des hétérocycles de nitrogène ou comprennent des fonctions de nitro-, amino-, ou carbamate. Un système automatique d'examen des échantillons de routine pour ces cinq paramètres serait donc extrêmement utile, nous permettant de séparer des échantillons dépassant les niveaux de tolérance et ensuite d'examiner ces derniers, si nécessaire, d'une manière plus complète et définitive.

Deux procédés fondamentaux—la détermination complète du phosphore joint au carbone et de l'inhibition de cholinesterase—ont été développés et sont discutés ici afin de démontrer la flexibilité et les promesses offertes par le domaine de méthodes automatiques en existence, telles que partition par solvant, hydrolyse, oxidation destructive, dialyse, formation d'une molécule colorée et colorimétrie. On y discute également en détail une méthode exacte et rapide de déceler automatiquement la présence du biphenyl dans le zeste d'agrumes afin de démontrer qu'il serait possible de développer de telles méthodes pour des résidus d'insecticides spécifiques; cette application comprend l'homogénéisation et l'extraction du zeste, la distillation par la vapeur, la partition par solvant, la partition par réaction au solvant et la spectrophotométrie aux rayons ultra-violets.

Zusammenfassung*

Pesticid-Rückstände in unserer totalen Umgebung sind überall von gesetzlichem und moralischem Interesse und in steigendem Masse auch eine Angelegenheit von staatlichen Dienststellen. Gewaltig ausgedehnte Bemühungen, um auf einer kontinuierlichen Basis Natur und Ausmass aller Arten von Pesticid-Rückständen in dieser Umgebung—einschliesslich Luft, Kleidung, Lebensmittel, Boden und Wasser—festzusetzen, haben das Problem

---

* Übersetzt von A. SCHUMANN.

von rückstandsanalytischer Unsicherheit in der erforderlichen Mikrogramm-Grössenordnung betont sowie den Mangel an ausgebildeten Rückstandsanalytikern. Ersteres ist eine Frage des erforderlichen Personals und der Laborausrüstung und ergibt sich aus der grossen Mannigfaltigkeit der vorhandenen Verbindungen und Substrate sowie aus der Kompliziertheit der buchstäblich hunderte von technischen Einzelheiten, die für den besonderen Zweck ausnutzbar sind. Letzteres ist eine Frage der Zeit und der Gründung von angemessenen Ausbildungszentren überall in der Welt, um Chemiker auf dieses Arbeitsfeld zu ziehen.

Beide Probleme können zu einem grossen Teil umgangen werden durch Automation von wichtigen und grundlegenden rückstandsanalytischen Arbeitsgängen. Zum Beispiel von den mehr als 900 vorhandenen pesticidwirksamen Verbindungen sind mindestens 400 bedeutende landwirtschaftliche Schädlingsbekämpfungsmittel. Die Hälfte von diesen enthalten organisch gebundenes Chlor; und ungefähr ein Drittel sind phosphororganische Verbindungen, welche ausserdem Cholinesterase hemmen. Einige dieser Verbindungen enthalten auch organisch gebundenen Schwefel; und viele Pesticide sind heterocyclische Stickstoffverbindungen oder enthalten Nitro-, Amino- oder Carbamatgruppen. Automatische, aussortierende Arbeitsgänge, um Routineproben auf diese 5 Parameter hin zu untersuchen, würden daher ungeheuer nützlich sein, um Proben, die möglicherweise über der Toleranz liegen, abzusondern für eine andere, genaue und endgültige Untersuchung, falls nötig.

Zwei Grundarbeitsgänge, Bestimmung von organisch gebundenem Phosphor und Prüfung auf Cholinesterasehemmung, sind entwickelt worden und werden erörtert, um die Biegsamkeit und Aussichten auf diesem Gebiet der vorhandenen automatischen Schritt-Methodiken zu veranschaulichen, welche Lösungsmittelverteilung, Hydrolyse, zerstörende Oxidation, Dialyse, Farbkörperbildung und Kolorimetrie einschliesst. Ausserdem wird eine schnelle und genaue automatische Sortiermethode für Biphenyl in Zitrusfruchtschale im einzelnen diskutiert, um Möglichkeiten zur Entwicklung dieser Methodik für spezifische Pesticid-Rückstände zu veranschaulichen; diese Anwendung schliesst ein automatische Homogenisation und Extraktion der Schale, Wasserdampfdestillation, Ausschüttelung mit Lösungsmittel, Ausschüttelung mit Säure zur Zerstörung von Pflanzeninhaltsstoffen und Ultraviolettspektroskopie.

References

BENDER, G. D.: Personal communication, August 1965.

BERAN, F.: Das Rückstandsproblem in Österreich. Residue Reviews 4, 82 (1963).

BLINN, R. C., and F. A. GUNTHER: Indirect ultra-violet spectrophotometric determination of sulphur dioxide by means of plumbous ion. Analyst 1027, 675 (1961).

—— —— The utilization of infrared and ultraviolet spectrophotometric procedures for assay of pesticide residues. Residue Reviews 2, 99 (1963).

—— —— The promising utility of infrared assay of pesticides and their residues. I. Stanford Research Institute Pesticide Research Bull. 2(3), 1 (1962); II. Ibid. 2(4), 3 (1962); III. Ibid. 3(1), 5 (1963).

BRO-RASMUSSEN, F.: Use and control of pesticides in Denmark. J. Assoc. Official Agr. Chemists 48, 29 (1965).

BRUAUX, P., S. DORMAL, and G. THOMAS: Détection et identification des résidus de pesticides organo-phosphorés par électrophorèse en gélose. Ann. Biol. Clin. 22, 375 (1964).

CALDERBANK, A.: Use of ion-exchange resins in residue analysis. Residue Reviews 12, 14 (1966).

CARMAN, G. E.: Personal communication, October 1965.

CASIDA, J. E.: Esterase inhibitors as pesticides. Science 146, 1011 (1964).

DORMAL, S., and H. HURTIG: Principles for the establishment of pesticide residue tolerances. Residue Reviews 1, 140 (1962).

FREEMAN, O. L.: As quoted in Environmental Health Letter 4(20), August 1, 1965.

FREHSE, H.: Special features in the analysis of pesticide residues: Residue analysis and food control. Residue Reviews 5, 1 (1964).

GAJAN, R. J.: Applications of polarography for the detection and determination of pesticides and their residues. Residue Reviews 5, 80 (1964); Recent developments in the detection and determination of pesticides and their residues by oscillographic polarography. Ibid. 6, 75 (1964).

GERKE, J. R.: Doctoral dissertation, Rutgers University, September 1964.

GUNTHER, F. A.: Instrumentation in pesticide residue determinations. Adv. Pest Control Research 5, 191 (1962).

—, and R. C. BLINN: Analysis of insecticides and acaricides. New York-London: Interscience 1955.

— —, and J. H. BARKLEY: Procedure for routine determination of biphenyl and o-phenylphenol on and in citrus fruit. Analyst 88, 36 (1963).

—, T. J. MILLER, and T. E. JENKINS: Continuous chloride-ion combustion method applied to determination of organochlorine insecticide residues. Anal. Chem. 37, 1386 (1965).

—, and D. E. OTT: Rapid automated determination of biphenyl in citrus fruit rind. Analyst (in press 1966).

HARRIS, T. H., and J. G. CUMMINGS: Enforcement of the federal Insecticide, Fungicide, and Rodenticide Act in the United States. Residue Reviews 6, 104 (1964).

KIRK, J. K.: Statement before the Agriculture Committee, U. S. House of Representatives, Washington, D. C., May 26, 1964.

MARTIN, II.: Present safeguards in Great Britain against pesticide residues and hazards. Residue Reviews 4, 17 (1963).

MEDVED', L. I., E. I. SPYNU, and IU. S. KAGAN: The method of conditioned reflexes in toxicology and its application for determining the toxicity of small quantities of pesticides. Residue Reviews 6, 42 (1964).

METCALF, R. L.: Personal communication, 1965.

MILLER, E. J.: The Pesticides Safety Precautions Scheme. Residue Reviews 11, 100 (1965).

MORRIS, W. W., JR., and E. O. HAENNI: Infrared spectra of pesticides. J. Assoc. Official Agr. Chemists 46, 964 (1963).

NATIONAL ACADEMY OF SCIENCES-NATIONAL RESEARCH COUNCIL: Pesticide Residues Committee, report on "no residues" and "zero tolerance." Washington, D. C., June 1965.

NETHERLANDS LEGISLATION: Netherlands Government Gazette (translated title) No. 66, April 5, 1965.

OTT, D. E., and F. A. GUNTHER: Procedure for the analysis of technical grade parathion in water plants by an anticholinesterase (AutoAnalyzer) method. J. Econ. Entomol. 59, 227 (1966a).

— — Rapid Screening for anticholinesterase insecticide residues by automated analysis. J. Assoc. Official Agr. Chemists. In press (1966b).

— — Automated analysis of anticholinesterase organophosphorus compounds on thin-layer scrapings. J. Assoc. Official Agr. Chemists. In press (1966c).

RAJZMAN, A.: Les résidus de biphényle dans les agrumes. Residue Reviews 8, 1 (1965).

SCHECHTER, M. S., and I. HORNSTEIN: Chemical analysis of pesticide residues. Adv. Pest Control Research 1, 353 (1957).

STRANDJORD, P. E., and K. J. CLAYSON: Automatic methods for the determination of serum lactate dehydrogenase and ornithine carbamoyl transferase activity. Technicon Internat. Symposium, New York, 1964.

SUTHERLAND, G. L.: Residue analytical limit of detectability. Residue Reviews 10, 85 (1965).

SUZUKI, I.: Japanese laws and regulations concerned with pesticide and food-additive residues in foodstuffs. Residue Reviews 4, 9 (1963).

WEINSTEIN, L. H., R. F. BOZARTH, C. A. PORTER, R. H. MANDL, and B. G. TWEEDY: Automated analysis of phosphorus-containing compounds in biological materials. I. A quantitative procedure. Contrib. Boyce Thompson Inst. 22, 389 (1964).

WESTENBERG, L.: The analysis of emulsifiable parathion preparations. Mededel. Landbouwhogeschool Wageningen 19, 554 (1954).

WILSON, C. W., and W. E. BAIER: Toward an equitable basis for assignment of residue tolerance values. Residue Reviews 4, 1 (1963).

WINTER, G. D.: Automated enzymatic assay of organic phosphate pesticide residues. N. Y. Acad. Sci. 87, 875 (1960).

—, and A. FERRARI: Automatic wet chemical analysis as applied to pesticide residues. Residue Reviews 5, 139 (1964).

ZWEIG, G., ed.: Analytical methods for pesticides, plant growth regulators, and food additives. Vol. I-IV. New York: Academic Press 1963-1964.

Paper No. 1686, University of California Citrus Research Center and Agricultural Experiment Station, Riverside, California.

Biochemical and metabolic changes in plants induced by chlorophenoxy herbicides

By

DONALD PENNER* and FLOYD M. ASHTON*

Contents

I.	Introduction	39
II.	Carbohydrates	40
III.	Lipids	44
IV.	Nitrogen metabolism	45
V.	Organic acids	53
VI.	Ethylene	55
VII.	Alkaloids	55
VIII.	Steroids, aromatics, and other compounds	56
IX.	Vitamins	56
X.	Pigments	57
XI.	Minerals	58
XII.	Water relations	63
XIII.	Endogenous auxins	64
XIV.	Nucleic acids	66
XV.	Enzymes	69
XVI.	Respiration	75
XVII.	Photosynthesis	79
XVIII.	Relationship of chemical structure to activity	81
XIX.	Fate of herbicides in plants	83
XX.	Conclusions	91
Summary		93
Résumé		94
Zusammenfassung		95
References		96

I. Introduction

A discussion[1] of chlorophenoxy herbicide-induced biochemical and metabolic changes involves consideration of plant composition, metabolic proc-

---

* Department of Botany, University of California, Davis.
[1] The following abbreviations for particular herbicides and other chemicals and terms will be employed: 2,4-D (2,4-dichlorophenoxyacetic acid); 2,4,5-T (2,4,5-trichlorophenoxyacetic acid); MCPA (2-methyl-4-chlorophenoxyacetic acid); silvex [2-(2,4,5-trichlorophenoxy) propionic acid]; NAA (napthaleneacetic acid); IAA (indoleacetic acid); 2,4-DB [4-(2,4-dichlorophenoxy)butyric acid]; RNA (ribo-

esses and pathways, relationship of chemical structure to biological activity, and the ultimate fate of the herbicides. Reviews by BRIAN (1964), SKOOG (1951), VAN OVERBEEK (1961 a), WOODFORD *et al.* (1958), WORT (1954), and WORT (1964 a and b) discuss various aspects of the effects of chlorophenoxy herbicides on plant composition and metabolism. Discussions on the mode of action of these herbicides are contained in reviews by BURSTRÖM (1955), CARNS and ADDICOTT (1964), CRAFTS (1961), HILTON *et al.* (1963), and VAN OVERBEEK (1961 b, 1964). Investigations into the various aspects of the fate of the herbicide in the plant are relatively recent and have been reviewed by AUDUS (1961), CRAFTS (1964), FREED and MONT-GOMERY (1963), SHAW *et al.* (1960), and WAIN (1961).

Plant responses to application of chlorophenoxy herbicides are manifold. The interpretations of these responses are complicated by the large number of interacting variables contributing to the observed responses. Response at the cellular level depends on the concentration of the herbicide which is controlled by its penetration into the plant, its subsequent transport, and differential accumulation in the plant (BLACKMAN 1950). The nature and physiological state of the plant, as well as the environment, play an important role in mediating the observed plant response.

The objectives of this review are to discuss the chlorophenoxy herbicide-induced biochemical and metabolic changes in plants and to integrate the interpretation of these responses into a comprehensive and meaningful picture. This review, although extensive, is not exhaustive and the reader is referred to the afore cited reviews for a complete survey of the literature.

The first part of the review will be devoted to a discussion of the effects of chlorophenoxy herbicides on plant constituents.

II. Carbohydrates

From an examination of the changes in carbohydrate content of chlorophenoxy herbicide-treated plants reported in the literature, it soon became apparent that the observations reported in part reflect the concentration of herbicide employed as well as the plant tissue analyzed. MITCHELL and BROWN (1945) reported that a 1,000 p.p.m. foliar application of 2,4-D was lethal to annual morning glory. Soon after the herbicide treatment the content of sugars had increased above the amount present in the control plants; it subsequently decreased until it was nearly depleted three weeks after the herbicide treatment. Starch and dextrin were also essentially depleted at this time. Their observations pointed out that observed change in plant constituents

nucleic acid); DNA (deoxyribonucleic acid); ATP (adenosine triphosphate); ADP (adenosine diphosphate); AMP (adenosine monophosphate); NAD (nicotinamide-adenine-dinucleotide); NADPH$_2$ (reduced nicotinamide-adenine-dinucleotide phosphate); CoA (coenzyme A); PPP (pentose phosphate pathway); SH groups (sulfhydryl groups); MCPB [4-(2-methyl-4-chlorophenoxy) butyric acid]; MCPP [2-(2-methyl-4-chlorophenoxy) propionic acid]; 4-CPA (4-chlorophenoxyacetic acid); and 2,4-DP [2-(2,4-dichlorophenoxy) propionic acid].

may be mediated by the time interval elapsed between herbicide treatment and time of plant analysis. This may help explain some of the apparently contradictory results.

RHODES (1952) followed the effects of MCPA on tomatoes over a 14-day period following the foliar application of the herbicide. He found no evidence for actual depletion of carbohydrates. The MCPA application prevented net synthesis of starch but did not reduce the net synthesis of sucrose. The sucrose yield of sugar cane has been shown to be increased by 2,4-D application (BEAUCHAMP 1951, CHACRAVARTI et al. 1955, CHACRAVÁRTI et al. 1956). However, this increase was not observed by CRUZADO and MUZIK (1950).

Application of 0.5 lb./acre of 2,4-D to Red McClure potato plants at early bloom stage resulted in increased specific gravity of the tubers, which is indicative of a higher starch content (PAYNE et al. 1953). NYLUND (1956) was unable to obtain this effect in Pontiac potatoes with either 2,4-D or MCPA. PAYNE and FULTS (1955) employed early combination treatments of maleic hydrazide and 2,4-D. This treatment caused a significant decrease in reducing sugars in the tubers, whereas late treatments caused a significant increase. Sucrose levels appeared to be unaltered. GRUODIENE (1963) sprayed the potato tubers before seeding and observed an increase in the starch yield of 23.7 percent in the subsequent crop over the control plants. CHAO and WANG (1957) showed that applications of 100 to 200 p.p.m. of 2,4,5-T failed to affect the content of starch and soluble sugar in potato tubers.

FREIBERG (1955) found that 2,4-D treated bananas were higher in reducing sugar content than the controls. Increased starch hydrolysis during the ripening of 2,4-D treated bananas has been shown by MITCHELL and MARTH (1944) and FREIBERG (1955). Apricots sprayed with 100 p.p.m. of 2,4,5-T were similarly shown to have fruits with higher contents of reducing sugar and sucrose than the controls (CRANE et al. 1956). YAKUSHKINA and KRAVTSOVA (1957) sprayed the flowers of peppers, tomatoes, and egg plants with a solution of 10 p.p.m. of 2,4-D. The 2,4-D treatment raised the sugar content of the fruits. SKRIPITSYNA (1950) showed that in the presence of 2,4-D, tomato fruits do not show the formation of phytin that accompanies usual seed formation. The seedless fruits accumulated sugars in the ovaries. In contrast to these results TUKEY and HAMNER (1949) observed that cherry fruits had a lower sugar content following fall application of a mixture of 100 p.p.m. of NAA and 16 p.p.m. of 2,4-D. OTANI (1959) reported that the preharvest spraying of persimmons with 2,4,5-T lowered the soluble tannin content. The soluble pectin content of a homogenate prepared from pea stem sections was doubled by 2,4-D treatment (GALSTON et al. 1963).

The general shift from starch to sucrose observed in fruits following applications of 2,4-D or 2,4,5-T associated with faster ripening has not been found in root storage tissue. Following applications of 2,4-D, EGOROVA et al. (1962), ZOSCHKE (1957), and RIRIE et al. (1952) observed reduced sugar contents in table beets and sugar beets. Carrot slices treated with 2,4-D also have shown decreased sucrose content (SAID and NAGUIB 1955). Non-storage roots such as onion tips have been reported by EGOROVA et al. (1962) to have an enhanced sugar content following application of granular 2,4-D.

CLAEYS (1950) followed the changes in the food reserves of germinating peas as a result of treatment with ammonium 2,4-dichlorophenoxyacetate. Following 24 hours of 2,4-D treatment the treated germinating peas showed an increased dry weight over the controls. During the early stages of germination these peas had a higher content of reducing sugars and sucrose. During late stages of germination the content of reducing sugars fell below the controls. At the time of primary root emergence the sugar content was lower, at the time of plumule emergence it was higher, and in the more developed seedling the sugar content again dropped. Starch-dextrin contents varied inversely with the sucrose contents. Apparently 2,4-D caused enhanced soluble sugar contents at the expense of reserve carbohydrates. Another aspect of the interaction between a chlorophenoxy herbicide and the carbohydrate level was investigated by Weaver (1961). He observed that the endogenous carbohydrate level present in the pea plant at the time of spraying was associated with changes in the tolerance of peas to MCPA.

The effects of chlorophenoxy herbicides on the carbohydrate contents of plant tissues seem to depend on which particular part of the plant is analyzed, the age of the plant, and the plant species, as well as which carbohydrate has been examined. LADONIN (1960) sprayed corn, wheat, sunflower, bindweed, and beans with the butyl and acetyl esters of 2,4-D and analyzed the plants from one to 15 days later. Only bindweed and beans were affected by the herbicide treatment. The carbohydrate content of the tops doubled, but that of the roots greatly decreased. In stems and leaves of 2,4-D sensitive plants, HOFMAN and V. SCHMELING (1953) and KASPERLIK (1955) have observed decreases in high molecular weight carbohydrates with increases in monoses or reducing sugars. In the roots the content of all examined carbohydrates decreased.

MONSTVILAITE (1962) noted that following 2,4-D applications on spring grain crops the content of starch in wheat and barley decreased by one to four percent whereas in oats it increased two to seven percent. ERGLE and DUNLAP (1949) showed that in cotton plants treated in the prebloom stage with 2,4-D the concentrations of sucrose, hemicellulose, and cellulose tended to increase in the leaves of the main stalk with increased amounts of 2,4-D applied. Reducing sugars varied inversely with the amount of 2,4-D applied. Leaf starch appeared to be affected only by the application of large amounts of 2,4-D.

Rapid increases in total sugars in leaves, stems, and underground parts of bindweed were reported by SMITH et al. (1947) following 2,4-D applications. These increases were followed by a return to the levels found in the controls. The starch-dextrin fraction decreased in all plant parts, and by the tenth day following 2,4-D treatment the level was only 1/3 the level found in the control plants. In a similar examination of the effects of 2,4-D on red kidney beans SMITH (1948) reported that four hours after 2,4-D application there was a decrease in the soluble sugars and starch-dextrins in the leaves. Three days after treatment the soluble sugar content had increased, while the starch-dextrins greatly decreased in the leaves of treated plants. In the stems both the soluble sugars and the starch-dextrins decreased. By the seventh day after the 2,4-D treatment there were only 46 percent of the soluble sugars and seven percent of the starch-dextrins in the leaves of treated plants compared to the controls. The stems from treated plants contained 21 percent of the soluble sugar and six percent of the starch-dextrin observed in control plants.

In *Sinapis alba* KOULA and KRAMLOVA (1964) found that MCPA and 2,4-D esters increased the reducing sugar and sucrose contents. GUNAR *et al.* (1952) showed that foliar applications of 2,4-D to sunflower caused an increase in the soluble carbohydrate content and an increase in the ratio of reducing sugars to sucrose. Studying the effects of 2,4-D treatment on inulin-storing plant tissue, WAIN *et al.* (1964) observed an increase in reducing sugars, but a 70 percent decrease in total carbohydrates. Insoluble carbohydrate levels remained unchanged; however, the tri- and higher water soluble polysaccharide content decreased. Following 2,4-D applications to wheat there appeared to be an increase in the soluble sugars in the leaves and an increased transfer of carbohydrate from leaves and stems to the spikes (TITOVA and MIKHAILOVA 1961). WOODBRIDGE (1962) noted that 2,4-D tended to cause an increase in the sugar level in the leaves and branches but a decrease in the sugar level in the roots of Bartlett pear. MACLEOD (1964) showed that *Vicia faba* leaves sprayed with a 5,000 p.p.m. MCPA solution showed a four-fold increase in fructose and glucose in the leaves. However, he also observed that when leaves of *V. faba* were detached, they too showed this build-up of sugars after three to four days. This led MACLEOD (1964) to suggest that the build-up of sugars caused by MCPA might be due to an interruption of their translocation from the leaves. SMITH *et al.* (1947) suggested that 2,4-D might well interfere with phloem function.

Soybean plants receiving 20 p.p.m. of 2,4-D in the nutrient solution *via* the roots were smaller, contained higher percentages of reducing sugars and hemicellulose, and had a lower percentage of starch than did untreated plants (WOLF *et al.* 1950). 2,4-D applied to the roots of *Lens culinaris* and the stems of *Cirsium arvense* resulted in a reduction in starch in the treated portion of the plant according to PILET and WURGLER (1953).

When wheat seeds were pre-treated with a ten p.p.m. solution of 2,4-D and the subsequent grain yield measured, BHARDWAJ (1962) observed a reduced carbohydrate content in the grain formed.

TUKEY *et al.* (1945) applied a 1,000 p.p.m. solution of 2,4-D to the leaves of bindweed plants. They observed a depletion of starch from almost all parts of the plant, but little hydrolysis of starch occurred in the chloroplasts. In some plant parts the loss of starch was correlated with active cell division. Starch hydrolysis appeared to be inhibited *in vitro* by 2,4-D. In contrast to this, OKUNTSEV and ZYRYANOVA (1959) stated that in their experiment the plant hydrolytic and oxidizing systems were highly activated. Furthermore, 2,4-D strongly stimulated plant respiration and caused a decrease in the sugar content. Carbohydrate reserves, hydrolyzable polysaccharides, reducing sugars, and non-reducing sugars were decreased in the stems of red kidney bean plants following 2,4-D applications (SELL et al. 1949). In the leaves and roots of 2,4-D treated red kidney beans SELL et al. (1957) obtained similar results with the exception that the polysaccharide content increased in the treated plants.

Chemical analyses by ELWELL and WEBSTER (1958) of twigs from oaks 70 days after they were sprayed with 2,4,5-T and silvex showed that 2,4,5-T reduced total sugars 73 percent and that silvex reduced total sugars 75 percent. MALISAUSKIENE (1961) treated a resistant and a susceptible variety of oats with 2,4-D at the rate of one kg/hectare during several stages of growth. In both varieties the concentrations of soluble sugars were reduced in all stages of growth measured. However, the resistant variety showed an increased concentration of starch as a result of the 2,4-D treatment. The content of glucose-1-phosphate in tomato buds declined following foliar treatment with 2,4-D (YAKUSHKINA 1956).

The effects of chlorophenoxy herbicides on the metabolic systems responsible for the altered carbohydrate composition of herbicide treated plants will be discussed in the section on respiration.

III. Lipids

The investigations of SELL et al. (1957) applying 2,4-D to red kidney beans show a reduction in the ether extract (lipids) in all plant parts except the stems of treated plants. In the stems SELL et al. (1949) found a slightly higher content of ether extract, unsaponifiable material, and fatty acids of the ether extract.

The application of 2,4-D to oat plants at the four-to-five leaf stage increased the kernel oil content (KENT and HUTCHINSON 1957). This effect was enhanced with increased 2,4-D dosage. Analyses by ZOSCHKE (1957) of sugar beets treated with 2,4-D and MCPA showed a reduction in lipid content following the herbicide treatment. He noted similar effects of 2,4-D and MCPA on the lipid content of flax. From the detailed studies of DUNHAM (1951) it is apparent that the application of four to 24 ounces/acre of 2,4-D to flax resulted in reduced oil content, up to 2.37 percent for the Minerva variety. A reduction in the iodine number also occurred. The maximum reduction in oil content occurred when the flax was sprayed at the pre-bud and at the late bud stages. Cotton seed oil content has been shown by EPPS (1953) to be reduced by 2,4-D application greater than 0.01 lb./acre.

The influence of chlorophenoxy herbicide on lipase activity have been examined by RAVAZZONI and VALERIO (1956). Olive oil hydrolysis by a lipase preparation from Ricinus was sharply inhibited by a concentration of 0.01 percent of 2,4-D. Similar reductions in lipase activity were observed by HAGAN *et al.* (1949). KUNERT (1959) observed that increasing concentration of 2,4-D inhibited the lipase activity in *Aspergillus niger*, whereas MCPA acid stimulated lipase activity.

The observed reductions of lipid content in the lipid storage areas of the plant seem to parallel the reduction of storage carbohydrates in carbohydrate storage areas of the plant. These changes would seem to be indicative of either a mobilization of storage products or reduction in the synthesis of storage products such as starch and lipids.

IV. Nitrogen metabolism

a) Total nitrogen

Chlorophenoxy herbicides appear to affect the accumulation, degradation, and movement of nitrogenous compounds in a general and in some instances a very specific manner. With the reversion of specific plant parts or organs to a meristematic state, there appears to be an increase in the total nitrogen level in these plant parts often at the expense of nitrogen from other parts of the plants as shown by the following investigations.

Analyses of twigs from oaks sprayed with 2,4,5-T and silvex made by ELWELL and WEBSTER (1958) 70 days after herbicide application showed an increase in the total nitrogen by 35 and 37 percent respectively. KENT and HUTCHINSON (1957) reported that 2,4-D applied to oat plants at the four-to-five leaf stage increased the kernel nitrogen content with increased dosages of 2,4-D. Fifteen days after tomato plants were sprayed with a ten p.p.m. solution of 2,4-D, YAKUSHKINA (1949) observed lower nitrogen levels in the leaves and a higher nitrogen content in the fruit. He concluded that 2,4-D only affected the redistribution of nutrients in the plant. The nitrogen content of excised corn roots was reduced only slightly at high 2,4-D concentrations according to KANDLER (1953).

The addition of three to 48 p.p.m. of 2,4-D to the nutrient media supplied to *Lemna minor* enhanced the nitrogen content on either a dry weight or a fresh weight basis (BLACKMAN 1956). HARRISON *et al.* (1958) stated that application of 2,4-D to potatoes resulted in a lowered nitrogen content in the juice extracted from the potato tubers.

As early as 1947 SMITH *et al.* (1947) observed that the total nitrogen in 2,4-D treated bindweed decreased in the leaves and increased in the stems, roots, and underground rhizomes. In 1949 WORT (1949) reported that in buckwheat treated with 1,000 p.p.m. of the sodium amine of 2,4-D the total nitrogen in the stems and roots increased and that of the leaves decreased four to eight days after the 2,4-D application. TORCHINSKAYA (1958 a) reported a loss of nitrogen from the leaves to the roots and stems of a number of weeds in which he followed the effects of 2,4-D applications.

The level of nitrogenous compounds in a plant part or organ may also vary with the period of time that has elapsed since the 2,4-D application.

SMITH (1948) showed that four hours after the application of two drops of 0.1 percent 2,4-D to each of the primary leaves of red kidney bean there was a decrease in the total nitrogen content of the stem above that of the controls and by seven days after 2,4-D treatment the total nitrogen in the stem was 4.6 times greater than that of the controls. Further investigations by REBSTOCK *et al.* (1952) showed that the greatest portion of this increase in the total nitrogen content in the stem was due to an increase in the insoluble nitrogenous fraction. They concluded that in the stems of 2,4-D treated plants, carbohydrates are utilized to synthesize the extra protein in the stem. Six days after applying a drop of 0.0054M 2,4-D or 2,4,5-T to the base of the blade of a primary bean leaf, SELL *et al.* (1957) reported finding an increase in the nitrogen content of all plant parts (i.e., roots, stems, and leaves).

In LADONIN's (1962) study on the influence of several herbicides on the nitrogen content of plants he noted that 15 days after 2,4-D applications the total nitrogen content in the leaves was reduced and the total nitrogen content in the stems was increased in wheat and kidney beans. However, in corn he observed an increase in the nitrogen content in the leaves and stems as compared to the controls. In field bindweed he found the opposite to be true, and the decrease in total nitrogen became greater with increasing time.

b) Nitrate

The addition of four p.p.m. of 2,4-D to the nutrient solution supplied to the roots caused soybean plants to show a decrease in total nitrogen, including nitrates, in the leaves and an increase in the stems and roots. FREIBERG and CLARK (1952) further found that all of the plant organs showed an increase in all soluble organic forms of nitrogen over the controls. More recently they showed that 2,4-D supplied in the nutrient media to nitrogen deficient soybean plants failed to alter the nitrate assimilation.

It appears that not only the physiological state of a plant, but also the nature of the plant influences the effect 2,4-D might have on the nitrate metabolism in the plant. SAID and YOUSSEF (1955) found that 2,4-D in the media induced rapid assimilation of nitrate nitrogen in radish-root slices, but suppressed the utilization of amino nitrogen. In 2,4-D sensitive sowthistle YAKUSHKINA and LIKHOLAT (1964) noted an accumulation of nitrate following application of one kg. of 2,4-D/hectare. This effect was not observed in 2,4-D resistant wheat.

Earlier studies by STAHLER and WHITEHEAD (1950) provided evidence that the leaves of 2,4-D treated sugar beets contained large enough quantities of potassium nitrate to be toxic to cattle. Extensive investigation by BERG and MCELROY (1953) indicated that following applications of two to eight ounces of 2,4-D/acre moderate increases in nitrate content in Canada thistle, dandelion, lambs quarters, redroot pigweed, Russian pigweed, and

Russian thistle were apparent. However, no increases in nitrate content were observed in the clovers, brome grass, timothy, alfalfa, and oats. FULTS et al. (1952) found only negligible increase in potassium nitrate content in potato plants sprayed at early bloom with 0.5 lb. of sodium salt of 2,4-D/acre. Field grown plants of corn and cucumber were sprayed with 2,4-D at various ages and analyzed at time intervals by BEEVERS et al. (1963): sprayed corn plants showed increased nitrate reductase, whereas sprayed cucumber plants showed decreased nitrate reductase activity.

In field studies SWANSON and SHAW (1954) found that 2,4-D sprayed sudan grass contained higher levels of nitrate shortly after herbicide application. But with time the levels of nitrate fell below that of the controls and then rose to the levels present or greater than those present in the controls by one month after the 2,4-D application.

c) Hydrogen cyanide

The hydrogen cyanide content of the sudan grass was influenced by 2,4-D applications (SWANSON and SHAW 1954). In the field the cyanide content of the grass was initially depressed; however, it rose rapidly, until four days after the 2,4-D treatment it was significantly higher than the cyanide content in the controls. GRIGSBY and BALL (1952) found no significant increase in the hydrogen cyanide content of wild black cherry (*Prunus serotina*) leaves following 2,4-D or 2,4,5-T applications. Similar applications of 2,4-D or 2,4,5-T were observed by LYNN and BARRONS (1952) to lower the cyanide content of wild pin cherry (*P. pennsylvanica*) leaves.

d) Amino acids and protein

Numerous investigators have attempted to determine the effect of 2,4-D application on the levels of protein and the various amino acids present in plant tissues. The interpretation of the obtained results may not be as contradictory as is at first apparent if one bears in mind that different concentrations of chlorophenoxy herbicides may have differing effects on the activity of the enzymes responsible for the observed levels of protein and amino acids and that some of the observed results may be due to interruption in translocation as suggested by MACLEOD (1964) or due to secondary effects such as the onset of death as suggested by MUZIK and LAWRENCE (1959). In their experiments they observed that uprooted bean plants left to die on the greenhouse bench underwent the same rapid decrease in root protein and amino acid content as their 2,4-D treated bean plants (MUZIK and LAWRENCE 1958 and 1959, LAWRENCE and MUZIK 1962).

MCPA at 5,000 p.p.m. sprayed on *Vicia faba* leaves caused a five-fold increase in the concentration of leucine, isoleucine, and valine in the leaves (MACLEOD 1964); when he observed unsprayed detached leaves he obtained the same results, leading him to suggest that MCPA might be interrupting the translocation of metabolites from the leaves and that this build-up might be a contributing factor in the toxic action of this herbicide.

Studying the influence of 2,4-D on free amino acid metabolism in carrot root tissue cultures, MÉNORET and MOREL (1958) observed an increase in the free amino nitrogen for the first 12 hours after their cultures received 2,4-D. After the initial rise, the free amino nitrogen level fell to about the original level after two days. The relative proportions of amino acids remained unchanged. Supplementing the growth media for potato tuber tissue cultures with 2,4-D ranging in concentrations from 10^{-3} to $10^{-7}M$ decreased the amino acid content in the experiments of FALUDI and FALUDI-DANIEL (1958). They concluded that increased deamination caused by 2,4-D was probably responsible for the reduction in the amino acid content. When PAYNE et al. (1951 and 1952) applied 2,4-D to the foliage of Red McClure potato plants they noted an increase in glutamic acid and a decrease in 11 other amino acids in the tubers harvested from the sprayed plants. TOMIZAWA and KOIKE (1954) reported that 2,4-D caused an increase in the glutamic and aspartic acid content of rice seeds. Following the treatment of onions with 2,4-D, KECK and HOFFMAN-OSTENHOF (1954) detected an increase in arginine-rich substances.

When 50 µg. of 2,4-D were applied to the primary leaf of a bean plant, AKERS and FANG (1956) observed a decrease in the total amount and in percent of aspartic and glutamic acid three and seven days after treatment. Feeding the 2,4-D treated plants $C^{14}O_2$ gave a three-to-four fold increase in the incorporation of labeled C^{14} into aspartic and glutamic acid over the control plants. Therefore, a very rapid synthesis of these amino acids appeared to be occurring, but their degradation must have been even more rapid. BUTTS and FANG (1956) also observed a greater incorporation of $C^{14}O_2$ into the amino acids of treated plants compared to non-treated plants.

SELL et al. (1949) applied a 1,000 p.p.m. 2,4-D solution to the base of the primary leaves of red kidney bean plants. The plants were harvested six days later and stem tissues were analyzed. The amino acid content was determined microbiologically. The stems of treated plants contained approximately twice as much protein as the stems of the non-treated plants. The same trend was observed for leucine, isoleucine, valine, phenylalanine, histidine, arginine, and threonine contents. The lysine and methionine contents of the stem were approximately three times that of the control. The tryptophan and aspartic acid contents were only slightly greater in treated than in non-treated plants.

SELL et al. (1949) and REBSTOCK et al. (1952) suggested that the decrease in carbohydrates and the increase in amino acid and protein contents in the stem tissue were due to the conversion in part of the carbohydrates in the stem into protein. WELLER et al. (1950) concluded that a part of the increase in the protein content of the stem following 2,4-D treatment may be accounted for by the translocation of nitrogenous compounds to the actively proliferating stem. They observed decreases in the amount of protein and of certain amino acids based on a dry weight basis in the leaves and roots of 2,4-D treated red kidney beans. RHODES (1952) concluded that a large proportion of the carbohydrates formed in the tomato were utilized for the synthesis of amino acids and proteins following foliar MCPA applications.

Foliar applications of 1,000 p.p.m. of sodium salt of 2,4-D caused significant increases of total free amino acids in sugar beet and potato tops, and a significant decrease in the free amino acid content of pinto bean tops (FULTS and PAYNE 1956). An interesting point is that potatoes and sugar beets have underground storage organs whereas pinto beans do not. FULTS and PAYNE (1956) also observed that the protein content increased in treated potato tops, stayed the same in treated sugar beet tops and decreased in treated pinto bean tops. 2,4-D affected the character of the protein in treated potatoes and pinto beans, but not in sugarbeets as measured electrophoretically. They suggested that the qualitative changes in the protein fraction of 2,4-D treated plants may be the fundamental basis for its selective action. Earlier PAYNE et al. (1953) had observed that a 0.5 lb./acre 2,4-D spray treatment on Red McClure potatoes increased the protein content of the tubers. YASUDA et al. (1955) noted that the protein content of potato tubers at harvest and after 60 days storage increased following foliar application of 0.5 lb./acre of 2,4-D to the potato plants. There also appeared to be a shift in the quality of the proteins due to the foliar 2,4-D application. Following the patterns of leaf proteins in Vicia faba and Raphanus sativus, by paper electrophoresis KONDO et al. (1957) noted a similar change in protein quality. The binding of 2,4-D to proteins as suggested by FREED et al. (1961 b) could account for the observed changes in protein quality.

Five days after the foliar application of 2,4-D on sugar beets, LIVINGSTON et al. (1954) observed a relative reduction in glutamine, alamine, lysine, and tyrosine content in the roots of the treated sugar beets. However, 60 days after 2,4-D applications to the foliage, the content of all free amino acids except asparatic acid had increased. EGOROVA et al. (1962) stated that 2,4-D treatment lowered the protein content of table beets. ZOSCHKE's (1957) studies showed an increase in protein content following 2,4-D treatment in wheat, barley, greenhouse grown flax, and sugar beets.

WORT (1951) showed that the protein nitrogen increased in the roots and stems of 2,4-D treated buckwheat plants; however, no change in protein nitrogen in the leaves was observed. As little as five p.p.m. of 2,4-D supplied to the roots of soybean plants increased the protein nitrogen in the roots and stems three days later, but decreased the protein nitrogen content in the leaves (FREIBERG and CLARK 1955).

When 2,4-D was applied at the usual rates for weed control in wheat, oats, and barley at the most tolerant stage of growth, no changes in protein content were noted. However, if the 2,4-D applications were made during susceptible stages of small grain growth, the yield was reduced and the protein content of the grain increased (SHAW et al. 1955). Using 2,4-D and 2,4,5-T, KLINGMAN (1953) noted similar results with winter wheat. Prior to this ERICKSON *et al.* (1948) had observed that wheat sprayed with 0.6 to 4.6 lb./acre of 2,4-D produced grain of a higher protein content. The protein content in the wheat varieties examined increased in direct relation to the amount of 2,4-D applied. HILL (1964) found that spraying hard red spring wheat for weed control did not materially alter the protein content of the wheat. Similar observations have been made by SHELLENBERGER *et al.* (1950) for hard red winter wheat. ALABUSHEV (1962) credited the control of weeds in barley, corn, and millet grains by 2,4-D applications, as being responsible for the observed increase in the per cent protein nitrogen in the grain following 2,4-D applications. The protein content of wheat, barley and oats were increased by one to two percent following the application of two kg. of 2,4-D/hectare by MONSTVILAITE (1962). 2,4-D applied at the rate of one kg./hectare to a resistant and susceptible variety of oats increased the protein content in the susceptible variety (MALISAUSKIENE 1961).

The results of FURTICK (1958) show that under field conditions, 2,4-D did not influence the protein content or dry matter yield of birdsfoot trefoil, ladino clover, tall fescue, and orchard grass. MEREZHINSKII (1962) found that 2,4-D had little visual effect on corn; however, it did lower the protein nitrogen content.

VOROB'EV and CH'A (1960) sprayed corn, peas, and *Rumex* sp. with 2,4-D and observed changes in protein, amino acid, and ammonia contents. In corn they observed that 2,4-D applications slightly lowered the protein content but notably increased the content of ammonia and amino acids, especially alanine, glutamine, glutamic acid, aspartic acid, glycine, and cystine. In peas, 2,4-D applications sharply lowered the protein content and increased the ammonia and amino acid content, especially alanine, glutamine, glutamic acid, glycine, and asparagine. In *Rumex* sp. the amino acid content was reduced 40 percent by 2,4-D applications.

Early applications of eight and 16 p.p.m. 2,4-D by WOODBRIDGE and KAMAL (1962) and KAMAL (1960) to Bartlett pears brought about changes in nitrogen metabolism the year of application. An increase in total and protein nitrogen in leaf tissue but a decrease in their content in stem tissue was observed. There was no apparent effect on the amino nitrogen in leaf or stem tissues. The treatment had variable effects on ammonia and amide nitrogen in both young and old leaves.

GRUODIENE (1963) sprayed potato tubers before seeding with 25 p.p.m. of 2,4,5-T. In the subsequent harvest the total nitrogen content increased two-fold and the protein nitrogen increased 1.5-fold relative to the control.

Treatment of bean and buckwheat plants with 2,4-D strongly depressed the rate of S^{35} incorporation into leaf proteins as demonstrated by VOROB'EV (1959). This effect was 1.5 times greater in bean leaves than in buckwheat leaves. In the case of buckwheat plants, the synthesis of leaf protein proceeded at an adequately high rate regardless of the inhibitive effects of 2,4-D. VOROB'EV and ABUEVA (1960) examined five-to-eight cm. high flax plants two days after they had been sprayed with 0.75 kg./hectare of the triethanol-amine salt of 2,4-D and found a reduction in the total nitrogen and protein nitrogen content. Ten days after the 2,4-D treatment, little protein synthesis had occurred and non-protein nitrogen predominated as free amino acids, especially asparagine. In plants 10-to-12 cm. high only temporary disturbances followed 2,4-D application. Ten days after treatment the treated plants had grown more rapidly than the controls but the protein nitrogen content was unaltered.

A compound that accumulated in soybean seedlings treated with 2,4-D was identified by KEY and WOLD (1961) as ascorbic acid. The increase in ascorbic acid concentration was associated with an increase in the concentration of protein sulfhydryl, trichloroacetic acid-soluble sulfhydryl, and reduced pyridine nucleotide groups.

The changes in protein content that occurred with senescence in excised cotton cotyledons were observed by BASLER and NAKAZAWA (1961) to be prevented by treatment with 2,4-D. OSBORNE and HALLAWAY (1961) showed that local applications of 2,4-D to leaves of *Euonymus japonica* caused an increase in the amounts of ethanol-soluble and -insoluble nitrogenous compounds in the treated area and a decrease in the quantities of these compounds in non-treated areas. The 2,4-D treated areas maintained an abnormally high rate of respiration. Carbon compounds were observed to move into or be preferentially retained in the treated areas of the leaf. Thus the 2,4-D treated area appeared to act as a metabolic "sink" as evidenced by the premature senescence of the untreated part of the leaf. No net protein breakdown occurred in the 2,4-D treated areas. Apparently 2,4-D caused the amino acid pool to increase in size and an increase in protein synthesis. The investigators concluded that 2,4-D was thus effective in delaying certain degradative processes of leaf aging, perhaps by acting in a stimulatory manner on the basic metabolic rate of the cells.

Further investigation in this area by OSBORNE and HALLAWAY (1960 and 1964) showed that when 25 µg. of butyl ester of 2,4-D were applied to the lamina of detached *Prunus* leaves, the treated areas remained green but the non-treated areas turned yellow after 11 days. Analysis for total nitrogen and 80 percent ethanol-insoluble nitrogen showed that nearly 50 percent of the protein was degraded in the control and untreated areas with a corresponding increase in free amino group, amides, and ammonia. The protein synthesis level in the treated area was maintained when checked by measuring the incorporation of C^{14} leucine. The peak of oxygen uptake in the detached leaves corresponded to the period of rapid protein degradation. The effectiveness of 2,4-D in maintaining protein levels in leaves increased as the leaves became older and this could be correlated with the decrease in their available endogenous auxin. Added 2,4-D appeared to influence net protein synthesis only when endogenous auxin levels fell to a level, which with other physiological factors was rate-limiting for protein synthesis.

WEST et al. (1960) found that in cucumber hypocotyl tissue, prior foliar application of 400 p.p.m. of 2,4-D increased the protein and RNA level above that of the controls. They concluded that the action of 2,4-D may be more directly concerned with RNA synthesis than with protein synthesis. This will be discussed in greater detail later.

The changes in protein and amino acid levels that have been observed may not only be due to the influence of 2,4-D on protein synthesis but also on the effects of 2,4-D on protein degradation. RAKITIN and ZEMSKAYA (1958) found that in beans treated with 2,4-D protein synthesis was decreased, the degradation of protein was enhanced as evidenced by increased protease activity, and there was increased movement of nitrogenous substances from the leaves to the roots. The same treatment produced only temporary changes in the nitrogen metabolism of oats. They attributed the resistance of the oats to 2,4-D to the differences in nitrogen metabolism in the two plant species. However, differences in herbicide concentration within the cells of the two species could make such comparisons more complex than is at first evident.

GUNAR et al. (1952) reported that 2,4-D caused a decline in nucleoprotein and phosphatide content and lowering of the protein nitrogen to nonprotein nitrogen ratio. LADONIN (1960) failed to observe this in sunflower, but did report (1960 and 1962) that 2,4-D lowered the protein content of the leaves and roots of beans and bindweed. A simultaneous increase in non-protein nitrogen was observed in these plant parts. That 2,4-D caused an irreversible decrease in protein accumulation and a simultaneous increase in non-protein nitrogen content in wheat seedlings, was demonstrated by TITOVA (1961). TORCHINSKAYA (1958 a) has reported increased protein hydrolysis in the leaves of a number of weeds and a loss of the nitrogenous compounds from the leaves to the roots and the stems. In another report (TORCHINSKAYA 1958 b), he found that 2,4-D infiltrated into the plant tissue caused a retardation of protein hydrolysis in sprouting lupine and in wilting makhorka tobacco leaves. The amino acids formed were deaminated and the liberated ammonia accumulated in the plant tissues, it was not utilized in amide synthesis. A decrease in the content of phosphate groups of ribonuclear proteins and an increase of free phosphate groups has been obtained by ELSAKOVA (1964) in maize, peas, and potatoes following 2,4-D treatment.

Soybean plants grown in a nutrient solution containing four p.p.m. of 2,4-D showed increased hydrolysis of leaf proteins with the subsequent translocation of the products to the stems and roots where additional protein was synthesized, according to FREIBERG and CLARK (1952).

In a later report FREIBERG and CLARK (1955) correlated proteolytic enzyme activity in the plant tissues of 2,4-D treated soybean plants with protein nitrogen contents. A root treatment of five p.p.m. of 2,4-D reduced the protein nitrogen content and the proteinase and peptidase activity in the leaves. The same treatment increased the protein nitrogen content, and the proteinase and peptidase activity in the roots and stems. They employed a 12 percent gelatin substrate for proteinase activity measurements and a six percent peptone substrate for peptidase activity measurements.

REBSTOCK et al. (1952) isolated a proteolytic enzyme preparation from 2,4-D treated red kidney beans. Enzyme activity was examined by measuring the increase in amino nitrogen utilizing a hemoglobin substrate. An increase in protein nitrogen in the stems of treated plants was correlated with an increase in proteolytic activity in the stems of almost 1/3 higher than in the non-treated plants. There was a decrease in proteolytic activity in the leaves and a very slight decrease in the activity in the roots. The decrease in proteolytic activity in the leaves was correlated with a decreased content of protein and amino acids in the leaves of treated plants by WELLER et al. (1950).

KEY (1964 a) found that a ten p.p.m. 2,4-D treatment increased the amount of leucine-C^{14} incorporated into soybean hypocotyl sections. Actinomycin D was able to inhibit this enhanced amino acid incorporation.

The difference in protein content of excised cotton cotyledons 48 hours after treatment with $10^{-8}M$ 2,4-D was shown to be due to changes undergone by the control and not by the treated cotyledons (BASLER and NAKAZAWA 1961).

GALSTON and KAUR (1959, 1961, and 1963) have shown that 2,4-D also affects the physical state of the cellular proteins. 2,4-D induced a decrease in the heat coagulability of the proteins of growing pea-stem cells; however, the total protein content was unaltered. The effect was most evident in the non-particulate phase of the cytoplasm. Auxin analogs which did not promote growth failed to have this influence or were less effective than those that did promote growth. This effect was not evident if 2,4-D or IAA were added in vitro. Ninety percent of the heat coagulum was protein, the rest nucleic acids, polysaccharides, and pectin. The auxin effect was inhibited by kinetin and ethionine. The effect of the latter could be reversed by methionine (GALSTON and KAUR 1963). The idea that the mechanism of action of a growth regulator such as 2,4-D might be an interaction of the growth regulating chemical with the enzymes of the protoplasm has received considerable attention. These interactions will be discussed in the section dealing with enzymes.

V. Organic acids

One of the first changes in the chemical composition of chlorophenoxy herbicide treated plants to be observed was the change in the total acid content. ERGLE and DUNLAP (1949) reported that the total organic acid content in cotton plants treated in the pre-bloom stage with 2,4-D varied inversely to the amount of 2,4-D applied. TUKEY and HAMNER (1949) applied a mixture containing 100 p.p.m. of NAA and 16 p.p.m. of 2,4-D to Prunus avuim and P. cerasus in October. This resulted in plant injury, light colored fruit, and a delay in maturity the following season. The sugar content of the fruit was lowered but the moisture content and the titratable acidity had increased.

When the leaves of field beans were pretreated with stimulatory levels of 2,4-D by HUFFAKER *et al.* (1962) several of the enzymes concerned in the dark fixation of carbon dioxide showed increased activity. Dark fixation resulted in the production of both sugar and large pools of organic acids in many plants. The acid content of 2,4-D induced parthenocarpic tomato fruits was shown to be unchanged by HWANG and SHING (1954). From their experiment on beans, SABUROVA and LOGINOVA (1961) concluded that stimulatory concentrations of 2,4-D increased the total acid content, whereas herbicidal dosages decreased the acid content. Furthermore the change in the metabolism of organic acids in beans was caused by the action of 2,4-D on the oxidative decomposition of carbohydrates.

The ascorbic acid and sugar content of onion root tips were shown by EGOROVA *et al.* (1962) to increase following applications of granulated 2,4-D. KEY and WOLD (1961) identified ascorbic acid as a compound that accumulated in soybean seedlings treated with 2,4-D. This accumulation was associated with an increased concentration of protein sulfhydryl, trichloroacetic acid-soluble sulfhydryl, and reduced pyridine nucleotide groups. Later KEY (1962) showed increases in the ascorbic acid content in the stems of cucumber seedlings, but decreases in the ascorbic acid content in the leaves following treatments with IAA, 2,4-D, and 2,4,5-T. The distribution pattern for dehydroascorbic acid was reversed from that of ascorbic acid. Increases of ascorbic acid in the stems were accompanied by a decrease in ascorbic acid oxidase activity and an increase in glucose-6-phosphate dehydrogenase activity. 2,4,5-T and *o*-chlorophenoxyacetic acid failed to produce this effect whereas silvex was active. Later Lin and Key (1964) found no significant connection between the effects of auxin on growth and the previously observed effects on the ascorbic acid-dehydroascorbic acid system.

MILLER and BURRIS (1951) noted that the oxidation of ascorbic acid was inhibited in a cell-free extract from barley by $1.2 \times 10^{-3} M$ 2,4-D or 2,4,5-T. The oxidation of glycolic acid was inhibited by a concentration of $10^{-4} M$ 2,4-5-T or MCPA. A crude bean-leaf juice extract was prepared by WAGENKNECHT *et al.* (1951) and the effect of 2,4,5-T and MCPA on ascorbic acid oxidase measured. Oxidation of ascorbic acid was inhibited 24 and 37 per cent by $0.0125 M$ 2,4,5-T and MCPA, respectively. Concentrations of the herbicides examined failed to stimulate ascorbic acid oxidase activity.

The amount of intracellular calcium oxalate has been shown by WASSBERG and GOODRICH (1956) to decrease in the leaves of *Datura stramonium* following 2,4-D applications. NANCE and CUNNINGHAM (1950) demonstrated that excised wheat roots placed in water containing $5 \times 10^{-3} M$ pyruvate showed increased production of acetaldehyde in the presence of $5 \times 10^{-4} M$ NAA. This effect was observed both under atmospheric conditions and in a nitrogen atmosphere.

The influence of chlorophenoxy herbicides on the tricarboxylic acid cycle has been studied by FALUDI and FALUDI-DANIEL (1958) and HOOS *et al.*

(1956). In potato-tissue cultures the total alpha-keto acid content increased with applications of 2,4-D allowing growth (10^{-7} and $10^{-5}M$), but decreased with 2,4-D concentrations which prevented growth ($10^{-3}M$) (FALUDI and FALUDI-DANIEL 1958). HOOS et al. (1956) observed that 2,4,5-T hastened ripening and improved size, color, and yield of fruit when sprayed on the trees at the pit-hardening stage of development. Both malic and citric acid contents decreased as the fruit ripened on the tree. Stimulation by 2,4,5-T seemed to be related to the faster rate of metabolism of these acids in the fruit.

The effects of chlorophenoxy herbicides on the TCA cycle will be considered in the discussion of the influence of these herbicides on plant respiration.

VI. Ethylene

The effects of 2,4-D on the ripening of Bartlett pears have been investigated by HANSEN (1946). Treatment of premature pears resulted in increased rates of ripening, respiration and ethylene production. The evolution of carbon dioxide was enhanced 1.3 times and that of ethylene 3.6 times over the controls. Mature fruits treated after harvest showed higher rates of respiration and ethylene production than did controls. Ethylene and 2,4-D in combination seemed to be more effective in enhancing ripening than either one separately. In contrast, MARTH and MITCHELL (1949) observed that 2,4-D reduced the effectiveness of naturally produced ethylene in bananas. However, applications of non-volatile forms of 2,4-D to bananas did not prevent the production of gaseous emanations capable of causing epinasty in the tomato plant.

HALL and MORGAN (1964) studying auxin-ethylene interrelations, showed that ethylene production was stimulated by 2,4-D in cotton plants sensitive to 2,4-D injury but not in sorghum plants which were insensitive to 2,4-D. Labeled ethylene was not produced from C^{14}-2,4-D. Ethylene stimulated IAA oxidase activity in cotton but not in the sorghum. This stimulation was most marked in the abscission zones.

VII. Alkaloids

Initial investigations on the influence of chlorophenoxy herbicides on the alkaloid contents of treated plants failed to show any effect of the herbicides on the alkaloid content. HOROWITZ (1949) noted that 2,4-D and silvex failed to affect the nicotine content of tobacco. TSAO and YOUNGKEN (1949) applied 2,4-D to Datura stramonium and failed to observe any significant alteration in the total alkaloid production. More recently SCIUCHETTI (1957) concluded from his research that 2,4-D possibly influenced changes in hyoscyamine/scopolamine ratios in D. stramonium.

Application rates of 2, 4 or eight lb./acre of silvex and 2,4,5-T caused increases in the total alkaloid contents of plants of tall larkspur, *Delphinium barbeyi* (WILLIAMS and CRONIN 1963). Silvex caused greater increases than did 2,4,5-T. WILLIAMS and CRONIN (1963) concluded that the quality of the alkaloids was unaltered, and only the quantity changed.

VIII. Steroids, aromatics, and other compounds

Repeated sprayings with 0.0001 percent 2,4-D by KADLUBOWSKI and SKALECKA (1961) increased the saponin (a steroid) content by 43 percent in soapwort and foxglove. The investigations of SORASUCHART *et al.* (1962) on the effect of 2,4-D on the glycoside content of *Ornithogalum umbellatum* showed that bulbs which received 2,4-D treatment had a higher mammilian lethal potency than the controls. Five glycosides were detected by chromatographic procedures. On hydrolysis they yielded the steroid strophanthidin.

Ayapin, scopolin, and scopoletin, all three coumarins, have been shown by DIETERMAN *et al.* (1964) to increase significantly in 2,4-D treated sunflower plants.

BARNES *et al.* (1962) found that applications of 2,4-D or gibberellic acid to the leaves or stems of *Hevea brasiliensis* seedlings increased the reduced thiol content and lowered the oxidized thiol content of the latex. Six days after 2,4-D treatment, WITSCH and KAPSERLIK (1955) observed that the aneurin content was usually depressed in the leaves while it increased in the stems and roots. It has also been shown by HOEPFNER (1961) that IAA and 2,4-D increased the inhibitor content in *Pisum sativum* roots as assayed by the *Lepidium sativum* and *Sinapsis alba* tests.

IX. Vitamins

Treatment of plants with chlorophenoxy herbicides appears to alter the vitamin content and perhaps the vitamin distribution within the plants. The growth of epicotyls of white lupine seedlings in a sterile culture media was not influenced by 2,4-D or 2,4,5-T; however, they did show an enhanced riboflavin content in experiments conducted by GUSTAFSON (1955). Flowers of peppers, tomatoes, and egg plants were sprayed with a 10 p.p.m. solution of 2,4-D by YAKUSHKINA and KRAVTSOVA (1957). The sprayed plants showed an increase in yield, in fruit sugar content, vitamin C content, and total dry matter. Applying 2,4-D and 2,4,5-T to tomatoes, FISHER (1961) observed similar results.

CATLIN (1959) studied the changes in some B vitamins associated with the responses of the Tilton apricot fruit to 2,4,5-T. Shortly after the 2,4,5-T treatment the concentrations of niacin, pantothenic acid, and biotin were observed to increase. Thiamine content was not altered until considerable growth differences had been established at which time the concentration in the treated fruits was lower than in the controls.

Following the application of 0.05 ml. of 0.1 percent 2,4-D to one of the primary leaves of the red kidney bean, LUECKE *et al.* (1949) noted the following changes in the vitamin content of various plant parts: the amounts of thiamine, riboflavin, and nicotinic acid decreased in the leaves and increased in the stems; the amount of pantothenic acid increased in both stems and leaves, whereas the amount of carotene decreased in both stems and leaves.

Changes in L-ascorbic acid have been discussed in the section dealing with organic acids.

X. Pigments

Applications of 2,4-D have been shown by a number of investigators (LADONIN 1960, MACIEJEWSKA-POTAPCZYKOWA 1953, MALISAUSKIENE 1949 and 1962, TORCHINSKAYA 1958) to reduce substantially the chlorophyll level in treated plants. Furthermore, MACIEJEWSKA-POTAPCZYKOWA (1953) showed that this change was accompanied by the appearance of pheophytin in the plant leaves.

The effects of chlorophenoxy herbicides on pigments other than chlorophyll have been investigated by SWANSON *et al.* (1956). Sections of Jerusalem artichoke tuber tissue in contact with filter paper impregnated with 2,4-D, 2,4,5-T or MCPA developed a red pigment. Germinating seeds of *Helianthus annuus* and *Taraxacum officinale* in contact with 2,4-D also produced a red pigment. Germinating seeds of *Lamium anplexicaule, Lepidium virginicum, Barbarea verna*, and *Agropyron pauciflorum* did not produce this red pigment.

Fruit ripening and associated color changes also appear to be influenced by chlorophenoxy herbicide treatment. WEINBERGER (1951) noted that foliarly applied 2,4,5-T at concentrations from ten to 40 p.p.m. advanced the maturity of peaches one to 17 days, but adversely affected size, color, shape, firmness, and quality of the fruit. Foliar applications of silvex applied two to four weeks before harvest have been shown to increase the intensity of the red color of Starking, Delicious, Turley, and Red Rome apples (BILLERBACK *et al.* 1953). Foliar application of 25 p.p.m. of 2,4,5-T plus ten p.p.m. of silvex caused an increase in red color of apples growing in the internal canopy without injury to foliage or reduction in fruit firmness (WHITE 1953).

Potato tubers of the Red McClure variety showed an intensified red color following 2,4-D treatments of the plants 69 days before harvest (FULTS and PAYNE 1955). NYLAND (1956) was able to show the same intensification of red color of the tubers of Pontiac potatoes following application of 0.5 to 1.0 lb. of 2,4-D/acre as a basal spray to plants in full bloom. The effects of 2,4-D on the pigments in the tomato fruit were shown to increase the carotene, lycopene, and xanthophyll contents (TAKAHASHI and NAKAYAMA 1960). The chlorophyll content of the fruit was decreased. Higher carotenoid levels have been observed in persimmon fruits following a preharvest spraying with the amine salt of silvex (OTANI 1959).

In certain instances 2,4,5-T has been observed to retard the loss of green color. HATTON (1958) found that 500 to 1,000 p.p.m. of 2,4,5-T proved effective in green color preservation of stored Persian lime fruits for four to eight weeks. The loss of green color by sepals of freshly cut heads of Waltham broccoli was retarded by dipping them into a 1,000 p.p.m. solution of 2,4,5-T (MARTH 1952).

XI. Minerals

The mobilization and redistribution of minerals following chlorophenoxy herbicide treatment appears to follow a pattern similar to that observed for amino acids, sugars, and certain vitamins. The mineral content of the leaves appears to decrease following herbicide treatment and that of the roots and stems appears to increase. The uptake of various ions may also be affected, but as WORT (1961) pointed out in his review, 2,4-D either depresses or has little effect on the uptake of some 14 ions by plants. In only a few instances is the ion uptake enhanced by herbicide treatment.

In all spring grain crops examined by MONSTVILAITE (1962) the ash content increased in direct proportion to the 2,4-D dose administered. However, SHELLENBERGER et al. (1950) reported that the mineral content of hard red winter wheat was not affected by 2,4-D spray applications. Some of the observed alterations in mineral content of chlorophenoxy herbicide treated plants may also have been due to the interference of the herbicide with electrolyte exudation by the plant. STALLWORTH (1962) has shown that 2,4-D treatment reduced the exudation of electrolytes by roots as well as excised stems of treated sweet corn plants.

Concentrations of 2,4-D greater than three p.p.m. in the nutrient media have been found by BLACKMAN (1956) to progressively depress the absorption of the total amount of nitrogen, phosphorus, and potassium by *Lemna minor* with both time and concentration. The greatest noticeable effect was on potassium uptake. However, on either a fresh or dry weight basis the contents of nitrogen and phosphorus were increased by three to 48 p.p.m. 2,4-D treatment. The addition of 20 p.p.m. of 2,4-D to the nutrient solution reduced the percentage of potassium and calcium in soybean plants (WOLF et al. 1950). The cation-to-anion ratio appeared to be unaffected by 2,4-D in later experiments (WOLF et al. 1952). The levels of phosphorus, sulfur and chlorine in the 2,4-D treated plants were inversely related to the nitrogen level present in the nutrient media.

WILDON et al. (1957) supplied 50 ml. of 0.1 percent sodium 2,4-D to the roots of young tobacco plants growing in sand culture. Two weeks later the treated plants contained less phosphorus and potassium in their tops than the controls. Calcium accumulated in the roots of treated plants. There was an increase in iron content and a decrease in sodium content in both the tops and roots of treated plants.

Top Crop bean plants were treated with a 100 p.p.m. solution of 2,4-D and the mineral uptake was compared with control plants by COOKE (1957). Within three to eight hours after 2,4-D application there was a marked increase in the uptake of K^{42}, Cl^{36}, Ca^{45}, and S^{35} by the treated plants. However, after 24 to 48 hours an inhibition was apparent, the uptake by the control plants of the labeled minerals was greater than by the 2,4-D treated plants. When RHODES et al. (1950) supplied 0.01 to 0.3 p.p.m. of MCPA to the roots of tomato plants, they observed a reduction in the potassium oxide content of the tops and lower stem and a marked accumulation of potassium in the roots. The potassium oxide content of the plant as a whole was greatly lowered. Apparently the 2,4-D treatment caused a redistribution of potassium in the plants, a specific inhibition of potassium uptake by the roots, and an interference in potassium transport. Applications of 2,4-D and 2,4,5-T to tomato plants increased the concentrations of manganese and copper in the fruits (FISHER 1961). When Jerusalem artichoke tuber tissue was pretreated with 2,4-D prior to the introduction of $Rb^{86}Cl$, HANSON and BONNER (1954) found the rate of Rb^{86} uptake was considerably reduced.

SWENSON and BURSTRÖM (1960) studied the simultaneous uptake of potassium, sodium, magnesium, and calcium by 2,4-D treated wheat plants. The 2,4-D treatment reduced uptake of these ions in the aforementioned order. They concluded that auxins do not specifically affect ion uptake, or act by virtue of their growth inhibition, but rather affect another cell property which is requisite for both cation uptake and water absorption. Working with woody plants, FEDEROV and EGOROVA (1963) showed that 2,4-D applications increased the P^{32} uptake by the bark and wood of the trunk, and wood of the roots. The uptake of calcium was generally inhibited by 2,4-D treatment.

The analysis of Cranberry bean plants by BASS et al. (1959) following 2,4-D treatment showed an increase in the percent potassium and sodium in the leaves of the treated plants. There was a decrease in the percent calcium but the level of the other elements analyzed remained almost the same as the control. In the stem tissue the percent of boron and potassium each increased but the percent of calcium, iron, and potassium each decreased with no significant change in the other elements examined. The roots of the treated plants showed an increase in the percent of boron, copper, potassium, and phosphorus, whereas the percent of magnesium, calcium, and sodium decreased. The amounts of iron and manganese present remained constant.

The investigations of SMITH (1962) and SMITH and HARRISON (1962) showed that 2,4-D caused a reduction in the calcium and phosphorus content of barley following pre-emergence applications of 2,4-D. Autoradiography showed that 2,4-D altered the Ca^{45} and P^{32} distribution in the treated plant. The Ca^{45} and P^{32} were concentrated in the proliferated lateral root areas of those plants treated with 2,4-D. The initiation of lateral root primordia from the pericycle was stimulated by 2,4-D.

RAKITIN and ZEMSKAYA (1958) stated that five ml. of 0.1 percent solution of 2,4-D/plant acted as a herbicide for beans but only slightly affected oats. In beans, the 2,4-D treatment stopped the uptake of nitrogen from the nutrient media, whereas in oats there were only temporary changes in the nitrogen metabolism.

When NANCE (1949) treated excised roots of four-day-old wheat seedlings with ten p.p.m. of 2,4-D, he observed an inhibition of nitrate absorption within three hours. The addition of one p.p.m. of potassium dihydrogen phosphate to the culture solution greatly increased this 2,4-D effect. YAKUSHKINA (1949) sprayed tomato plants with ten p.p.m. of 2,4-D solution and analyzed the treated plants 15 days later. He observed a lower nitrogen content in the leaves, an increase in phosphorus in the leaves and fruit, an increase in the ash content of the fruit, and a modest increase in the nitrogen content of the fruit. He concluded that 2,4-D only affected the redistribution of nutrients. In a later report (1956) he showed that the content of glucose-1-phosphate in buds of 2,4-D treated plants declined. RHODES (1952) showed that the uptake of nitrogen and phosphorus pentoxide was reduced in tomatoes 14 days after foliar application of MCPA. The MCPA treatment had no apparent effects on potassium, calcium, or magnesium uptake. Similarly BEREZOVSKII and KUROCHKINA (1957) showed that a foliar application of a 50 p.p.m. solution of 2,4-D on sunflowers lowered the P^{32} uptake from the soil and reduced the content of phosphorus/unit dry weight of the plant. Phosphorus tended to accumulate in the roots at the site of entry in the treated plants.

The uptake of P^{32} by cultured pieces of potato tissue was shown to be diminished by FALUDI (1962). Incorporation of phosphorus into compounds with high energy bonds was slightly inhibited, that into proteins and nucleic acids was slightly promoted, and that into phospholipids was markedly increased. He concluded that the excess lipid phosphorus came from the medium. In earlier reports FALUDI et al. (1959) and FALUDI and FALUDI-DANIEL (1960) observed that the cells even lost a significant part of their own phosphorus content. The equilibrium in the 2,4-D treated tissue shifted toward the decomposition of organic phosphorus compounds, especially that of the acid soluble and lipid soluble fractions.

In intact plants FANG and BUTTS (1954 a and b) found that 2,4-D reduced the upward movement of P^{32} to the leaves proportionately to the 2,4-D concentration applied. Apparently phosphorus uptake can be both enhanced or inhibited by 2,4-D treatments depending on the plant species. PARIS (1960) observed that 10^{-5} to $10^{-7}M$ 2,4-D inhibited the phosphorus uptake in *Triticum vulgare* and *Brassica campestris pleifers*, but increased the phosphorus uptake in *Pisum sativum*. Within 30 hours following a 2,4-D treatment, VLASYUK and STARCHENKOV (1962) noted an increase in the phosphorus uptake by corn plants. Similar treatment of sugar beets resulted in an accumulation of phosphorus in the plant with a retardation of the rate of phosphorus metabolism in organic fractions and an increase in the inorganic fraction. After several days the level of total phosphorus in the beet plant was reduced but its metabolism was accelerated under growth stimulatory conditions. After spraying two varieties of corn with various forms of 2,4-D, MEREZHINSKII (1962) found that these treatments raised the soluble phosphorus content of the treated corn plants.

2,4-D applied at herbicidal concentrations to 2,4-D susceptible plants has been shown to inhibit the entrance of phosphorus into the plant and certain subsequent phosphorus transformation in the plant (BEREZOVSKII and KUROCHKINA 1956). Over a period of time this inhibition appeared to be overcome by the plant to a degree. BEREZOVSKII and KUROCHKINA (1956) concluded that 2,4-D may affect the intermediate products of phosphorus metabolism probably as a result of the inhibition of synthetic processes in the primary stages of phosphorylation. The synthesis of more complex phosphorus compounds like nucleoproteins and phosphotides would also be inhibited. However, they stated that the changes in phosphorus metabolism cannot be regarded as the primary reaction of the plant to 2,4-D nor can the inhibition of phosphorus metabolism caused by 2,4-D be identified with the disturbance caused by phosphorus starvation.

A decrease in the uptake of P^{32} by soybean plants following 2,4-D treatment has been observed by TOMIZAWA (1956). The 2,4-D treatment increased the inorganic phosphorus content in all parts of the plants but decreased the incorporation of phosphorus into certain metabolic intermediates such as 3-phosphorylyceric acid, glucose-1-phosphate, and hexose-diphosphates. The increase of phosphorus in the stems of the 2,4-D treated plants appeared to be due to the movement of phosphorus from the leaves to the stems.

The phosphorus content in treated plants appeared to follow the same pattern that held true for proteins, amino acids, nucleic acids, phosphorus, and vitamins. REBSTOCK et al. (1954) observed that 1,000 p.p.m. of 2,4-D foliarly applied to Cranberry bean plants caused a decrease in the total amount of phosphorus in the treated plant. In the stems of treated plants the phosphorus content increased whereas in the leaves it decreased. The decrease in the leaves occurred in the acid- and alcohol-soluble fraction and to some extent in the nucleic acid fraction. In the stem, the acid soluble fraction had increased slightly, the alcohol- soluble fraction had not changed appreciably, but the phosphorus in the nucleic acid fraction had almost doubled. With the reversion of the stem tissue to a meristematic condition, such changes in phosphorus metabolism would be expected. FANG and BUTTS (1954 a and b) obtained results similar to REBSTOCK et al. (1954) and concluded from their radioautographs that the 2,4-D treated area acted as a metabolic sink. The percent of phosphorus assayed as inorganic phosphorus decreased in the stem and increased in the leaves of the treated bean plants compared to the nontreated plants indicating a higher percent of organic phosphorus compounds in the stem than in the leaves of treated plants. Their observation that the incorporation of P^{32} into glucose-1-phosphate was increased in the stem and decreased in the leaves of 2,4-D treated plants as compared to non-treated plants appeared to substantiate this. In their experiments with 2,4-D treated barley and bean plants, RAKITIN and POTAPOVA (1959 c) have also found decreases in the phosphorus content of young leaves. On the other hand, ROSS and SALISBURY (1962) noted increased hydrolyzable phosphate levels in beans and cocklebur plants after the leaves were dipped in 2,4-D solutions, whereas DNP treatment reduced the hydrolyzable phosphorus level.

Increased action of phosphate enzyme systems in 2,4-D treated tomato plants has been observed by RAVAZZONI (1949). Increased phosphatase activity and phosphate content in 2,4-D or 2,4,5-T treated rice seedlings and sweet potato plants have been reported by TOMIZAWA and KOIKE (1954).

Species of *Commelina* and *Xanthosoma* were analyzed by LOUSTALOT *et al.* (1953) 24 hours and one week after treatment with 2,4-D. The percentage of water-soluble phosphorus was consistently higher in treated plants than in untreated plants. 2,4-D treated white beans showed an increase in inorganic phosphate content in the leaves, stems, and roots 48 hours after treatment compared to untreated plants or uprooted plants left to die. One week after treatment, the inorganic phosphate content of the roots of the treated plants had increased markedly, however, there were no apparent increases in the phosphate content of the stems and leaves.

Applications of growth stimulating dosages of 2,4-D and 2,4,5-T to tomato flowers by RAKITIN *et al.* (1956) resulted in increased phosphorus content. The protein-lipid phosphate and the ratio of organic phosphorus to inorganic phosphorus increased. Growth inhibitory dosages of 2,4-D and 2,4,5-T resulted in an accumulation of phosphorus by the plant, primarily in the inorganic forms. The ratio of organic phosphorus to inorganic phosphorus was also observed to decline.

OMROD and WILLIAMS (1960) studied the effects of 2,4-D on the phosphorus metabolism of *Triflorium hirtum*. They reported that 21 day old plants sprayed with 50 µg. of 2,4-D showed a striking increase in acid-soluble organic phosphorus and a decrease in inorganic phosphorus within one minute after treatment. Treatments of five µg. were adequate to cause this effect in the petioles and stems. When 500 µg. of 2,4-D were applied to the leaves, there was a definite time lapse before the increase in acid-soluble organic phosphorus was observed, while in the petioles and the stems this dose reduced this fraction. From their results OMROD and WILLIAMS (1960) concluded that the action of 2,4-D was related to phosphorus metabolism.

The diversity of the observed effects of chlorophenoxy herbicides on the mineral composition of plants may be due to factors such as a differential effect between growth stimulatory concentration of the herbicide and herbicidal concentration, a differential response between plant species dependent on their genetic and physiological background, an effect on the transport of inorganic as well as organic materials, and a differential effect on a number of facets of inorganic ion metabolism.

XII. Water relations

The influence of chlorophenoxy herbicides on the water relation of treated plants is mediated by the concentration of herbicide applied, the plant species examined, and the plant organ examined. Fruits of Derby apricots sprayed with 2,4,5-T were shown by CRANE et al. (1956) to be higher in moisture content than unsprayed fruits. Increases in moisture content of cherry fruits have been shown by TUKEY and HAMNER (1949) following the application of a mixture containing 100 p.p.m. of NAA and 16 p.p.m. of 2,4-D.

BROWN (1946) observed that a 1,000 p.p.m. foliar 2,4-D application to bean seedlings increased the moisture content of the stems and the plant as a whole but decreased the moisture content of the leaves. However, low concentrations of 2,4-D were shown by FREIBERG and CLARK (1952) to increase the moisture content of soybean leaves.

The results obtained by WAIN et al. (1964) indicated that eight days after a 2,4-D treatment there was a 75 percent increase in the moisture content of disks from the storage tissues of Jerusalem artichoke and chicory. Increased water uptake by Jerusalem artichoke tissues following 2,4-D treatment has also been observed by HANSON and BONNER (1954). TOMIZAWA and KOIKE (1954) were unable to find any appreciable effect of 2,4-D on the moisture content of treated rice seedlings and sweet potato plants.

The leaves of *Pelergonium zonale, Parthenocissus quinquefolia, Onethera biennis, Chenopodium album,* and *Tropaeolum maius* were sprayed with a 0.1 percent solution of 2,4-D by MACIEJEWSKA-POTAPCZYKOWA (1957). The uptake of water was inhibited; however, evaporation increased. SWENSON and BURSTRÖM (1960) stated that 2,4-D treated wheat plants also showed reduced water uptake. The experiments of BASS-BECKING and EVERSON (1953) pointed out that in tissue preparations of *Bryophyllum calycinum* the effect of 2,4-D on subturgid cells seemed to be the lowering of the equilibrium volume or maximum turgidity in water. This effect was concomitant with an increase in permeability which was particularly noticeable in plasmalysis.

Part of the change in the water relations of chlorophenoxy herbicide treated plants appears to be the influence of the herbicides on the transpiration in treated plants. A reduction in transpiration of 34 percent five days after a 1,000 p.p.m. foliar 2,4-D application to bean seedlings was measured by BROWN (1946). RAKITIN and POTAPOVA (1959 a) reported that 2,4-D caused a drop in transpiration in sunflower and oat plant. The sunflower showed a decreased protoplasmic viscosity and a simultaneous increase in protoplasmic permeability following 2,4-D treatment. MCPA at 10^{-3} to $10^{-6}M$ was found by KOZINKA (1964) to reduce the transpiration rate of two-week-old barley seedlings whether it was applied to the leaves or supplied to the culture solution. Similar reductions in transpiration have been reported in wheat seedlings by YAMADA and MURATA (1950) following

2,4-D treatments. 2,4-D and 2,4,5-T were shown by STAN (1960) to cause similar reductions in the transpiration rate of tomatoes.

The nature of the plant species and the physiological state of the plant may mediate the influence of the chlorophenoxy herbicide on plant transpiration. The experimental results obtained by PLAYER (1950) showed that applications of a 0.01 or 0.02 percent 2,4-D spray to castor beans and corn reduced the daytime transpiration in the castor bean, but increased the nighttime transpiration. The transpiration of corn appeared to be unaffected. Transpiration increases of 0.5 to two times normal have been reported by MALISAUSKIENE (1959) for 2,4-D treated oat plants in the three-to-four leaf stage. POPOV et al. (1954) found that apricot and apple leaves sprayed with a 500 p.p.m. 2,4-D solution, to prolong spring dormancy as a protection against frost, showed a higher rate of transpiration than did the controls.

Chlorophenoxy herbicide treatments affected stomatal closure in the sensitive plants examined by KASPERLIK (1955). The transpiration decreased sharply and the water content increased following a 2,4-D treatment of sensitive plants. In oats, a relatively insensitive plant to 2,4-D, no such effects were observed. REPP (1954), ZELITCH (1961), and BRADBURY and ENNIS (1952) have also observed that the application of 2,4-D caused an inhibition of stomatal opening.

From the results of these investigations it is conceivable that the observed effects of chlorophenoxy herbicide treatment on transpiration and the subsequent water content of the plant are closely related to the effects of these herbicides on photosynthesis. A reduction in photosynthesis would result in the eventful closure of the stomata. This in return would explain a reduced transpiration intensity and increased water content in treated plants.

XIII. Endogenous auxins

An indoleacetic acid-forming enzyme system from pea plants studied by LIBBERT (1960) converts tryptophan to IAA; 2,4-D inhibited this reaction. MOEWUS (1954) traced hormone biosynthesis in the germinating zygote of the alga *Chlamydomonus eugametos*; 2,4-D at 0.01 p.p.m. stimulated the following deaminations: tryptophan → indole-3-propionic acid and indole-3-acrylic acid → phenylalanine.

There are implications that chlorophenoxy herbicides can replace endogenous auxins. This is exemplified by the experiments of CRANE (1954) in which Tilton apricot fruits, whose ovules had been killed by frost, grew to normal size when sprayed with 40 p.p.m. of 2,4,5-T two days after the frost occurred. Fruit development requires endogenous auxins which is produced by the developing ovule.

The translocation of IAA in 2,4-D treated bean seedling stems and roots was shown to be inhibited (HAY 1955 and 1956). ZWAR and RIJVEN (1956) noted that treatment of hypocotyl segments of *Phaseolus vulgaris* with 2,4-D,

2,4-DP, and 2,4-DB inhibited basipetal IAA transport. They concluded this may have been caused by competition for sites on hypothetical carriers of IAA.

Pretreatment of pea stem tissue with 2,4-D increased the amount of indole-3-acetic acid aspartate formed three to five times in an experiment performed by SUDI (1964). He examined other carboxylic acids possessing auxin activity, obtaining similar results; however, those carboxylic acids not possessing auxins activity failed to produce this effect.

The levels of free endogenous IAA in shoots, leaves, and roots of peas, beans, and sunflowers appear to be unaffected by 2,4-D treatment according to AUDUS and THRESH (1956 a and b). However, LOCKHART and WEIN-TRAUB (1957) report that 2,4-D caused a marked decrease of the free auxin content of the terminal shoots of bean seedlings, on both a concentration and an absolute basis. The decrease in auxin content preceded the decrease in rate of growth.

In addition to the effects of chlorophenoxy herbicides on auxin synthesis, transport, and complexing, evidence has accumulated which shows that these herbicides influence auxin degradation. FANG and BUTTS (1957) measured IAA destruction by following the release of $C^{14}O_2$ from C^{14}-IAA. Following 2,4-D treatment the rate of release of carbon dioxide from IAA in beans and corn was reduced. Oxidative decarboxylation of IAA by corn and pea shoots was increased in the light (FANG et al. 1959); however, light did not influence the destruction of IAA in the roots or in other tissues pretreated with a $10^{-4}M$ 2,4-D solution. ANDREAE and ANDREAE (1953) found that IAA was oxidized by a pea brei with the liberation of hydrogen peroxide. This oxidation was limited by a light-activated step. 2,4-D stimulated the IAA oxidation presumably by accelerating the light activated step. As early as 1949 GOLDACRE (1949) observed that 2,4-D increased the rate of IAA oxidation by a crude enzyme preparation isolated from etiolated pea epicotyl. The addition of boiled onion juice to the system inhibited IAA oxidation, but 2,4-D reversed this inhibition.

Young tissue from etiolated peas were found by GALSTON and DALBERG (1954) to be low in endogenous levels of IAA oxidase; however, as the tissue aged the level of IAA oxidase increased. 2,4-D or 2,4-DB infiltrated into the etiolated pea tissue increased the IAA oxidase content.

In vitro experiments of HENDERSON and DESSE (1954) and HENDERSON et al. (1954) showed that 2,4-D enhanced the rate of destruction of IAA in peas and sunflowers but decreased the rate of destruction in oats. In vivo experiments showed that pea and sunflower plants treated 2,4-D contain less auxin than untreated ones, while the opposite was true for oat plants.

It appears that changes in auxin level in a particular plant organ sub-sequential to chlorophenoxy herbicide application may be influenced by the effects of the herbicide on auxin synthesis, complexing, and degradation, as well as auxin transport.

XIV. Nucleic acids

The demonstration that auxin treatment caused an increase in the RNA and DNA contents of tobacco pith (SILBERGER and SKOOG 1953) gave rise to the hypothesis that the mechanism of auxin action was linked with nucleic acid metabolism (SKOOG 1954). Following 2,4-D treatment of onions, KECK and HOFFMAN-OSTENHOF (1954) detected an accumulation of thymonucleic acid and arginine-rich substances.

In soybean leaves, PERUANSKII (1964) found that DNA accumulated whereas the RNA content decreased following 2,4-D and 2,4,5-T applications. On the other hand in corn, MASHTAKOV et al. (1964) noted that corn plants treated with MCPA showed an increase in RNA and soluble nucleotide content. These increases were not associated with the usual growth stimulation but rather growth inhibition. From this MASHTAKOV et al. (1964) concluded that MCPA interfered with the biochemical function of the nucleic acids.

Cytochemical evidence presented by SATAROVA (1961) showed higher RNA contents of meristematic areas of gladiolus tissue from tissue cultures supplied a media containing 2,4-D. This led to the conclusion that 2,4-D participated in the regulation of nucleic acid metabolism. Prior to this, MASUDA (1959) had already concluded that the physiological effects of IAA involved a role played by RNA. In his experiments RNAase treated sections of Avena coleoptile did not show elongation following IAA treatments. There was a significant release of inorganic phosphorus, auxin, and some sugars during the RNAase treatment. When coleoptiles were pretreated with IAA before RNAase exposure they elongated without a lag when again treated with IAA.

KEY (1959) found that 24 to 48 hours after the application of $5 \times 10^{-4}M$ 2,4-D to 2.5-day-old soybeans there were increases in proteins, nucleic acid, and acid soluble nucleotide in the hypocotyls. Analysis of the soluble nucleotide fraction showed large increases in ATP. Later KEY and HANSON (1961) found that the protein/RNA ratio decreased in three-day-old soybeans for 48 hours after treatment. After that the ratio increased due to a decline in RNA. Increases in the following acid soluble nucleotides were also observed: cytidine mono- and triphosphates; adenosine mono-, di-, and triphosphates; guanosine mono- and triphosphates; and uridine mono-, di-, and triphosphates. KEY (1959) concluded that 2,4-D induces both promotive and inhibitory processes in the plant. The promotive processes are postulated to act through nucleotide metabolism, and inhibitory processes through the formation of natural growth inhibitors.

WEST et al. (1960) observed that 400 p.p.m. of 2,4-D foliarly applied to cucumber seedlings increased the amount of RNA present in the hypocotyl tissue of the treated plants. Since RNA plays a vital role in protein synthesis, WEST et al. (1960) concluded that the action of 2,4-D might be more directly concerned with RNA synthesis than with protein synthesis. OOTA

(1964) has recently reviewed RNA biosynthesis and its relation to protein synthesis.

Soybean plants harvested 48 hours after being sprayed with $5 \times 10^{-4}M$ 2,4-D contained twice as much RNA in the seedling hypocotyls with over half of the increase appearing in the microsomal fraction and a fourth in the soluble fraction (CHRISPEELS and HANSON 1962). Changes in nuclear RNA content were not measured specifically. An increase in synthesis of RNA and protein in the region of rapid cell proliferation is necessarily preceded by increased nuclear activity. Since directly or indirectly RNA and protein synthesis are controlled by the DNA, CHRISPEELS and HANSON (1962) believed that the primary site of action of 2,4-D could well be the nucleus. The cytochemical basis of 2,4-D action would then lie in renewed nuclear activity and the reversion of the tissue to meristematic metabolism.

KEY (1963) found that 2,4-D did not appreciably affect the incorporation of adenosine-8-C^{14} into RNA, the rate of adenosine-8-C^{14} incorporation appeared to be unaltered as did the quality of the RNA. Later KEY and SHANNON (1964) showed that concentrations of 2,4-D and IAA that promoted cell elongation, five to 25 p.p.m. of 2,4-D in the bathing solution for tissue slices, enhanced ADP-8-C^{14} and ATP-8-C^{14} incorporation into the RNA of excised soybean hypocotyls. Inhibitory levels, 100 to 500 p.p.m. of 2,4-D, decreased C^{14}-nucleotide incorporation. ADP incorporation into RNA could be inhibited by actinomycin D (KEY 1963). A net increase in RNA content, primarily ribosomal in fully elongated cells, of 25 to 30 percent following treatment with 25 p.p.m. of 2,4-D was observed by KEY and SHANNON (1964). They concluded that a net transfer of C^{14}-RNA from the nucleus to the ribosomes occurred under the influence of 25 p.p.m. of 2,4-D. Actinomycin D strongly inhibited RNA synthesis and prevented increases in ribosomal RNA. From the known action of actinomycin D (REICH et al. 1961), they concluded that the induced RNA synthesis occurred on the DNA template.

Employing actinomycin D, 8-azaquanine, and puromycin as inhibitors, KEY (1964 a) reported that the enhancement of cell elongation by 2,4-D required active RNA synthesis and in turn protein synthesis. He presumed that the rate of formation of some specific RNA is enhanced directly or indirectly by auxin, which leads to an increased supply of some limiting enzyme or enzyme system (i.e., protein synthesis). Later KEY (1964 b) stated that cell elongation is independent of ribosomal RNA synthesis and probably s-RNA synthesis. Auxin-induced elongation instead appeared to depend upon the synthesis of a fraction of RNA having the general properties of messenger RNA. In these experiments actinomycin D at low concentrations inhibited RNA synthesis but not auxin-induced growth.

GALSTON and PENNY (1964) showed that auxin-induced growth by etiolated pea epicotyl sections could be reduced or inhibited by one to ten p.p.m. of actinomycin D or about 100 p.p.m. of crystalline ribonuclease. In green pea epicotyl only actinomycin D had this inhibitory ability. These inhibitors failed to affect endogenous growth. Therefore, they closely linked auxin-induced growth to RNA metabolism.

Labeled messenger RNA has been isolated from green pea stem sections following short term (one to two hour) incubations with $10^{-5}M$ C^{14}-carboxyl labeled IAA (BENDANA et al. 1964). This labeling was inhibited by ten p.p.m. of actinomycin D. For both 2,4-D and IAA, the carboxyl label appeared more effective than the methylene label. The incorporation of the labeled auxin and carbon dioxide into RNA was stimulated by light indicating that some of the effect is due to decarboxylation and photosynthetic carbon dioxide recycling. The extent to which the labeled auxin is metabolized and the label subsequently incorporated into RNA is not known.

During a 12-hour incubation, ten p.p.m. of 2,4-D enhanced the metabolic breakdown of RNA in excised corn mesocotyl tissue whereas 800 p.p.m. of 2,4-D almost completely inhibited RNA breakdown after only four hours. This enhanced breakdown of RNA at low concentrations of 2,4-D occurred entirely at the expense of microsomal and soluble RNA, with no measurable change in the RNA content of the nucleus and the mitochondria (KEY 1963). Low concentrations of 2,4-D promoted loss of RNA by enhancing the metabolic breakdown of pre-existing and newly synthesized RNA as shown by adenine-C^{14} incorporation experiments. High concentrations of 2,4-D inhibited the loss of RNA, apparently by inhibiting degradation of RNA as observed four hours after treatment.

With excised corn mesocotyl tissue, concentrations of 2,4-D up to 50 p.p.m. were shown by SHANNON et al. (1964) to accelerate growth and RNAase activity in parallel. At higher 2,4-D concentrations both were inhibited. Protein content decreased throughout the experiment. In intact tissue, normal expansive growth of corn mesocotyl and cucumber hypocotyl was accompanied by an increase in RNAase activity (SHANNON 1963). 2,4-D promoted or inhibited growth depending upon the concentration of 2,4-D. Low levels of 2,4-D promoted growth and RNAase activity while herbicidal concentrations inhibited both. The development of RNAase activity in plant tissues examined by SHANNON (1963) appeared to parallel maturity rather than growth per se. High levels of 2,4-D were shown by SHANNON (1963) and SHANNON et al. (1964) to induce synthesis of nucleic acid and protein. The increased synthesis was proportional to the log of the 2,4-D concentration employed. SHANNON (1963) associated the increases in nucleic acids and proteins with increased cell division especially in the region of the vascular bundle in the cucumber hypocotyl. He concluded that the action of 2,4-D is mediated through the inhibition of normal maturation processes, possibly by blocking RNAase activity. Simultaneously it seems that the nucleus is induced to renewed activity giving rise to cell division, aberrant growth, and the ultimate death of the plant.

BASLER and NAKAZAWA (1961) noted that excised cotton cotyledons treated with $10^{-3}M$ 2,4-D did not undergo the changes in RNA content usually associated with senescence. Forty-eight hours after the 2,4-D treatment the differences in RNA content between the treated and control cotyledons were attributable to the loss of RNA in the control. The RNA content of the treated cotyledons appeared static and the RNA synthesizing machinery apparently unaffected. 2,4-D did not block RNA degradation in isolated microsomes; however, the addition of glutathione did block this degradation (BASLER 1962).

From later studies by BASLER and HANSEN (1964) it appears that both RNA synthesis and degradation are inhibited by relatively high concentrations of 2,4-D ($10^{-3}M$). Sucrose appeared to be necessary for increased RNA contents to occur in the particulate nucleic acid fraction during culture. Sucrose also stimulated the loss of nucleic acid in the RNA containing supernatant fraction. The investigators concluded from their labeling experiments that 2,4-D reduced the movement of labeled RNA from cell particulates to the cytoplasmic fraction or that 2,4-D caused preferential incorporation of C^{14}-orotic acid into DNA of the cell nucleus. The latter possibility might be due to the renewed nuclear activity following 2,4-D treatment postulated by CHRISPEELS and HANSON (1962).

Nucleotides in combination with 2,4-D or IAA were shown by WEDDING and BLACK (1964) to stimulate growth of pea epicotyl segments as measured by fresh weight increase during a 20-hour incubation. The most effective nucleotide was NADP.

XV. Enzymes

Contributing to the complexity of the biochemical and metabolic changes induced by chlorophenoxy herbicides are their effects on enzymes. Following herbicide applications, the cellular concentration of the herbicide capable of affecting enzyme activity or synthesis may vary between plant organs. The cellular concentration of the herbicide in a given organ can also vary with the plant species examined. Chlorophenoxy herbicides may further affect observed enzyme activity by influencing the physical state of the enzyme, the adsorption of enzymes to cell particulates, the availability of cofactors, and the molecular architecture of the enzyme itself. All of the aforementioned factors must be taken into consideration in the interpretation of the effects of herbicides on enzymes. Recent listings of the effects of herbicides on various enzymes have been compiled by WORT (1961 and 1964 a).

The effects of growth regulator herbicides on the physical state of enzymes has also received consideration (FREED et al. 1961 b). FREED et al. (1961 b) have shown that herbicides such as 2,4-D are capable of both stimulating and inhibiting enzymes depending upon the concentration of the herbicide. It has been shown by GLASZIOU (1952) that 2,4-D is effective in binding pectin methylesterase to cell-wall preparation from tobacco pith and Jerusalem artichoke tubers. The significance of such a system would be that the auxin-mediated binding of pectin methylesterase to the cell wall could possibly control expansive growth by controlling esterification of pectic substances. The activity curves presented by GLASZIOU (1957) are consistent with the hypothesis that auxin binds pectin methylesterase to the cell-wall receptor sites by an adsorption mechanism which can be ascribed by the Langmuir isotherm. The pectin methylesterase preparation from artichoke tubers was separated into three fractions by GLASZIOU and INGLIS (1958). The adsorption of one of these fractions to the cell-wall could be increased by 2,4-D. From his studies with intact disks from tobacco pith, GLASZIOU (1958) suggested that 2,4-D shifted the equilibrium position for partitioning the enzymes between the disks and the external medium by controlling the binding of the pectin methylesterase to sites in the cell-wall. MACEY (1960) reported that pectin methylesterase activity of artichoke tube tissue was increased by 2,4-D under conditions favorable to a physiological response to 2,4-D. Interaction of this effect with calcium indicated that a certain stabilized surface was involved in the experiments. General esterase activity was observed to be suppressed by high concentrations of 2,4-D or IAA, but not by nonchlorinated phenoxyacetic acids.

In the following discussion the effects of chlorophenoxy herbicides on the activities of a number of enzymes and possible mechanisms of their influence on enzyme will be considered. Pectin transeliminase has been isolated by ALBERSHEIM and KILLAS (1962) from an acetone powder prepared from pea seedlings. The amount of inhibition of this enzyme by IAA or 2,4-D appeared to be influenced by the concentration of the substrate. Increasing concentrations of pectin increased the degree of inhibition. The auxin antagonist 2,4-dichlorophenoxyisobutyric acid was as effective as 2,4-D. Later ALBERSHEIM (1963) showed that the $\Delta 4{:}5$ unsaturated uronide, the product of the action of pectin transeliminase on citrus pectin, was capable of showing product inhibition. Ozone removed this inhibition whereas 2,4-D and IAA enhanced the inhibition, but were not inhibiting in the absence of the product. ALBERSHEIM suggested that 2,4-D inhibited the action of this enzyme by formation of a triple complex among auxin, enzyme, and product. Pectin methoxylase (pectase) activity was shown to decrease in red kidney bean plants by NEELY et al. (1950 b) following 2,4-D applications.

NEELY et al. (1950 a) found that 2,4-D considerably lowered both the α- and β-amylase activity in the stems of treated red kidney bean plants. They were unable to detect α-amylase activity in the leaves of the treated or non-treated plants. An initial stimulation of amylase activity and then a fall in activity have been reported in rice by TOMIZAWA and KOIKE (1954) and RAVAZZONI (1951). WORT and COWIE (1953) obtained similar results with β-amylase in wheat. VASCUTANU et al. (1961) observed that low concentration of 2,4-D caused a stimulation of amylase activity in the treated plants whereas high concentrations caused an inhibition of enzyme activity. Increases in amylase activity following applications of chlorophenoxy herbicides have been reported by GASPARYAN and AVETISYAN (1958) in seeds of kyurushna grass and HOFMANN and v. SCHMELING (1953) in a number of weeds. The latter authors also reported enhanced activity of saccharase following 2,4-D treatments.

FREIBERG (1955) reported 2,4-D treated bananas to be higher in diastatic activity than controls four days after treatment. The effects of chlorophenoxy herbicides on a large number of enzymes concerned with carbohydrate metabolism have been examined and have even been found to cause shifts in the metabolic pathways employed by the treated plants (HUMPHREYS and DUGGAR 1961, BLACK 1960, BLACK and HUMPHREYS 1962). Increases in ascorbic acid in the stems of 2,4-D and 2,4,5-T treated cucumber seedlings were accompanied by a decrease in ascorbic acid oxidase and an increase in glucose-6-phosphate dehydrogenase activity in the studies of KEY (1962). *In vitro* studies by BLACK (1960) and BLACK and HUMPHREYS (1962) employing 2,4-D treated, etiolated three-day-old maize seedlings have shown increased utilization of ribose-5-phosphate, and an increased rate of oxidation of both glucose-6-phosphate and 6-phosphogluconate in enzyme extracts from 2,4-D treated tissues. Activities of glucose-6-phosphate dehydrogenase and 6-phosphogluconate dehydrogenase were higher in the treated seedlings. *In vitro* studies with the enzymes of the glycolytic pathway showed decreased activity of 6-phosphofructokinase, aldolase, and glyceraldehyde-3-phosphate dehydrogenase in extracts from 2,4-D treated corn seedlings. The activity of pyruvic kinase also decreased but the activities of phosphoglucoisomerase, phosphoglyceric kinase and enolase were not affected by 2,4-D pretreatments. The enhancement of the activity of the enzymes in the pentose phosphate pathway (PPP) and the inhibition of the enzymes in the glycolytic pathway led BLACK (1960) and BLACK and HUMPHREYS (1962) to conclude that 2,4-D initiated a shift from the glycolytic pathway to the PPP in 2,4-D treated etiolated corn seedlings.

FREED *et al.* (1961 b) studying crystalline enzymes noted a marked stimulation of the activities of glyceraldehyde-3-phosphate dehydrogenase, glucose-6-phosphate dehydrogenase, and isocitric dehydrogenase by low concentrations of 2,4-D, whereas high concentration of 2,4-D (1,000 p.p.m.) caused a decline in the activities of these enzymes. A decrease in the synthetic function of the TCA cycle following 2,4-D treatments has been suggested by HUMPHREYS and DUGGAR (1961).

Early studies of WEDDING and BLACK (1962) showed that $10^{-3}M$ 2,4-D uncoupled oxidative phosphorylation and inhibited oxidative activity of malic dehydrogenase. A 100-fold purified malic dehydrogenase has been prepared from red beet root mitochondria by WEDDING and BLACK (1963). The reaction catalyzed by this enzyme proceeding in either direction was inhibited by 2,4-D. Their results showed that the inhibition was competitive for oxidized or reduced NAD but noncompetitive for malate or oxaloacetate. They concluded that a possible mechanism might be through a nonenzyme complexing of 2,4-D with pyridine nucleotides, and the subsequent complex would then inhibit the malic dehydrogenase.

LEOPOLD and GUERNSEY (1953) proposed that auxins, 2,4-D, and 2,4,5-T might act in plant growth and metabolism at least in part by reacting with coenzyme A(CoA). They prepared a mitochondrial preparation from tomato ovaries and measured the disappearance of free SH groups from the added CoA. The greater the concentration of 2,4-D or 2,4,5-T present the greater the inhibition of the disappearance of the SH groups.

Invertase activity has been shown by TOMIZAWA and KOIKE (1954) to rise and then decline following 2,4-D treatment. They also noted that phosphatase activity in 2,4-D treated rice seedlings and sweet potato plants increased. WORT and COWIE (1953) had observed an initial increase in phosphatase and phosphorylase activity followed by a decrease in activity in Marquis wheat plants. The changes in phosphatase activity normally undergone by the senescence of excised cotton cotyledons were prevented by the addition of 2,4-D to the culture media (BASLER and NAKAZAWA 1961).

Spraying tomatoes with 2,4-D has been shown by YAKUSHKINA (1956) to increase phosphorylase activity, with an increase of synthetic activity in the buds, and an increase in hydrolytic activity in the leaves of treated plants. In contrast, 2,4-D appeared to noncompetitively inhibit potato phosphorylase in the experiments of LEARY and CLAGGETT (1951). Decreases in phosphorylase activity were reported by NEELY et al. (1950 b) in both stems and leaves of 2,4-D treated red kidney bean plants.

Wheat germ lipase was shown by KVAMME et al. (1949) to be noncompetitively inhibited by 2,4-D; however, the extent of the inhibition was only 1/400 of the 2,4-D inhibition of castor bean lipase reported by HAGEN et al. (1949). RAVAZZONI and VALERIA (1956) also found that olive oil hydrolysis by a lipase preparation from castor bean, *Ricinus*, was sharply inhibited by 2,4-D. Unlike these results, MOLOTKOVSKII and VOLOTOVSKAYA (1950) reported that 2,4-D caused a stimulation of lipase isolated from castor bean seeds. They concluded that 2,4-D might act by activation of lipase, which cleaves fatty acids from lecithins of the plasma membranes. This would presumably cause changes in microstructure and increase wall permeability to various nutrients.

Investigations employing cultures of *Aspergillus niger* by KUNERT (1959) showed that increasing concentrations of 2,4-D inhibited lipase activity, whereas MCPA stimulated lipase activity. Pretreatment of kyurushna grass seed with 0.005 percent 2,4-DB caused an increase in lipase activity (GASPARYAN and AVETISYAN 1958).

Enzymes involved in nitrogen metabolism are also influenced by 2,4-D treatments. High concentrations of 2,4-D were found by VASCAUTANU et al. (1961) to stimulate urease activity. BEEVERS et al. (1963) reported that 2,4-D treated corn seedlings showed increased nitrate reductase activity, whereas treated cucumber showed reduced reductase activity. The hydrolytic activity of proteases in beans has been observed by RAKITIN and ZEMSKAYA (1958) to increase following 2,4-D applications; similar increases were not observed in oats. A 24 hour addition of five p.p.m. of 2,4-D to the nutrient media significantly increased the level of proteinase and polypeptidase in the stems and decreased the levels of these enzymes in the leaves of soybean plants five days after the 2,4-D treatment (FREIBERG 1952, FREIBERG and CLARK 1955). REBSTOCK et al. (1952) applied 2,4-D to the primary leaf of red kidney bean plants and found increases in proteolytic enzyme activity in the stems and decreases in activity in the leaves and roots of treated plants.

Infiltration of 2,4-D or 2,4-DB into etiolated pea tissue caused an increase in indoleacetic acid oxidase content (GALSTON and DALBERG 1954). The influence of 2,4-D on other enzyme systems in auxin metabolism have been reported by LIBBERT (1960) and MOEWUS (1954).

Applications of 2,4-D and 2,4,5-T at stimulative growth rates to tomato flowers resulted in increased activity of oxidation-reduction enzymes (RAKITIN et al. (1956). Growth inhibiting dosages caused disorganization of the oxidation-reduction system. Increases in catalase activity following applications of chlorophenoxy herbicides have been reported by STAN (1960), SIVORI and CLAVER (1950), GRITSAENKO (1961), VASCAUTANU et al. (1961), GASPARYAN and AVETISYAN (1958), WORT and COWIE (1953), KOULA and KRAMLOVA (1964), BYNOV and TITOVA (1960), and LADONIN (1960). The increases in catalase activity detected by LADONIN (1960) were found in treated bindweed and beans but not in treated corn, wheat, or sunflower. BYNOV and TITOVA (1960) observed that the 2,4-D induced increase in catalase activity was much greater in the stems and roots than in the leaves of five-day-old wheat plants. Potato catalase was inactivated by ten to 1,000 p.p.m. of 2,4-D according to RAVAZZONI (1951).

Chlorophenoxy herbicides did not appear to alter the catalase content of wheat (LADONIN 1960, TSITOVICH et al. 1955, KOULA and KRAMLOVA 1964) or of castor bean (RAVAZZONI 1951). Decreases in catalase activity following chlorophenoxy herbicide applications have been reported by TSITOVICH et al. (1955), TOMIZAWA and KOIKE (1954), KOULA and KRAMLOVA (1964), and GRITSAENKO (1961).

Enhanced peroxidase activity following chlorophenoxy herbicide application have been observed by RAKITIN and POTAPOVA (1959), MOREL and DÉMETRIADES (1955), STAN (1960), SIVORI and CLAVER (1950), GASPARYAN and AVETISYAN (1958), RAVAZZONI (1951), and GRITSAENKO (1961). KOULA and KRAMLOVA (1964) reported both increases and decreases in peroxidase activity depending upon which chlorophenoxy herbicide and which formulation were applied. An initial increase followed by a decline in peroxidase activity were reported by WORT and COWIE (1953). GAGARINA (1964) applied 0.025 ml. of a 0.1 percent 2,4-D solution leaf to leaves of corn and beans. High peroxidase and normal flavine dehydrase activities were observed in the corn, whereas flavine dehydrase activity decreased markedly in the beans and peroxidase remained at a normal level. Peroxidase activity was unaltered in rice seedlings and sweet-potato plants following 2,4,5-T applications (TOMIZAWA and KOIKE 1954). Similarly

LADONIN (1960) found the peroxidase levels unaltered in corn, wheat, and sunflower after 2,4-D applications. Chlorophenoxy herbicide induced decreases in peroxidase activity have been reported by KOULA and KRAMLOVA (1964), LADONIN (1960), MEREZHINSKII (1962), and GRITSAENKO (1961).

Sunflower plants treated with 2,4-D showed lowered ascorbic acid oxidase activity; however, 2,4-D treatments during bud formation enhanced ascorbic acid oxidase activity (RAKITIN and POTAPOVA 1959 a). These results point out the importance of the physiological state of the plant in determining the observed results. Cytochrome oxidase activity in sunflower plants was also enhanced by 2,4-D treatment (RAKITIN and POTAPOVA 1959 a).

Polyphenol oxidase activity has also been shown to be enhanced by chlorophenoxy herbicides (RAKITIN and POTAPOVA 1959 b, STAN 1960). Reductions in polyphenol oxidase activity following applications of chlorophenoxy herbicides have been observed by KOULA and KRAMLOVA (1964), LADONIN (1960), MEREZHINSKII (1962), MOREL and DÉMETRIADES (1955), and GASPARYAN and AVETISYAN (1958).

This discussion of the effects of 2,4-D on plant enzymes is by no means exhaustive but only serves to give an indication of what has been observed. WORT (1954 and 1964 a) reviewed the different mechanisms by which 2,4-D might give the results observed by the many investigators in 2,4-D research. He suggested that it is only logical to infer that the processes concerned with synthesis of enzyme proteins might also be altered and result in altered molecular architecture. Species vary in their content of specific proteins, so that changes in the quality of protein could account for the variation in results between plant species.

An example of the problems encountered in the interpretations of herbicidal effects on enzymes as stated in the first paragraph of this section on enzymes are the effects of 2,4-D on proteolytic enzymes. Proteolytic enzymes are known to increase in activity in the stem and decrease in the leaves of treated plants (FREIBERG 1952, FREIBERG and CLARK 1955, REBSTOCK et al. 1952). One could easily assume that the activities of other enzymes might vary in a like manner. The amount and quality of proteins and amino acids are changed by 2,4-D treatment (YASUDA et al. 1956); perhaps the effect of 2,4-D on the activity of some enzymes may also be brought about by increased or decreased supply of vitamins for prosthetic or coenzyme function, or an altered supply of amino acids for synthesis. In other words the effect on these enzymes by 2,4-D treatment could be quite secondary in nature. LUECKE et al. (1949) have observed such changes in vitamin content of 2,4-D treated red kidney beans.

The effects of 2,4-D on plant enzymes thus far examined are generally considered secondary in nature.

XVI. Respiration

Stimulation of respiration in plant tissue following applications of chlorophenoxy herbicides have been reported by ALLEN (1953), BASLER and NAKAZAWA (1961), BROWN (1946), FANG et al. (1960 and 1961), KANDLER (1953), LADONIN (1960), LINDEN (1952), MALISAUSKIENE (1962), MAXIE and CRANE (1956), NAKAZAWA (1950), OKUNTSEV and ZYRYANOVA (1959), OSBORNE and HALLAWAY (1961 and 1964), RAKITIN and POTAPOVA (1959 b), SAID and YOUSSEF (1955), SMITH et al. (1947), SOUTHWICK (1946), STAN (1960), UOTA and DEWEY (1953), WAIN et al. (1964), WILLIAMS (1963), and YAMADA and MURATA (1950).

Chlorophenoxy herbicide-induced inhibition of respiration has been reported by FREELAND (1949), MITCHELL et al. (1949), TAYLOR (1947), WRIGHT (1961), JOHNSON and COLMER (1957 and 1958), MAXIE and CRANE (1956), WEST and HENDERSON (1950), and WORTH and McCABE (1948). NANCE (1949) apparently was unable to find any effect of 2,4-D on the oxygen uptake of excised wheat roots.

CATLIN and MAXIE (1959), SAID and NAGUIB (1955), and MALISAUS-KIENE (1959) have noted that chlorophenoxy herbicides may initially stimulate respiration; this stimulation is then followed by a decline in respiration rates.

Respiration appeared to be stimulated following applications of low to intermediate concentrations of 2,4-D, whereas high concentrations of 2,4-D inhibited respiration in the experiments of WEDDING et al. (1954) and KELLY and AVERY (1949 and 1951). The latter authors also noted that an 18 to 20 percent respiration stimulation in oat tissue required at least 1,000 times the 2,4-D concentration required to produce the same stimulation in peas (KELLY and AVERY 1949). In pea tissue greater respiration stimulation following 2,4-D treatment was observed in starved than in non-starved stem tissue (KELLY and AVERY 1951). The results of KELLY and AVERY (1949) show that dinitrophenol (0.1 to 1.0 p.p.m.) had a greater stimulatory effect on oat tissue respiration than 2,4-D (0.3 to 200 p.p.m.) and inhibition set in at a lower concentration.

WILLIAMS (1963) reported that respiration was stimulated by a 500 p.p.m. spray of 2,4-D under all light qualities, but significantly so only under red light. The presence of two and five percent sucrose in the 500 p.p.m. 2,4-D spray reduced the 2,4-D effect. The respiratory quotient was unaffected in the 2,4-D-induced respiration stimulation observed by KANDLER (1953) in excised corn roots, whereas in the respiration inhibition in wheat reported by TAYLOR (1947) there was an increase in the respiratory quotient. 2,4-D was found by OSBORNE and HALLAWAY (1964) to stimulate oxygen uptake in young and mature *Prunus* leaves, but have little effect on autumn leaves. The 2,4-D treated areas of *Euonymus* leaves maintained abnormally high respiration rates. Carbon compounds either moved into or were preferentially retained in these areas. OSBORNE and HALLAWAY (1961) concluded that the 2,4-D treated areas acted as metabolic sinks, as evidenced by the premature senescence of the untreated part of the leaf. They further suggested that 2,4-D might act by a stimulation of the basic metabolic rate of the cells.

Organisms capable of anerobic respiration were reported by WORTH and MCCABE (1948) not to be affected by 2,4-D, whereas the respiration of those organisms that required oxygen was inhibited. Inhibited oxygen uptake caused by 2,4-D has been overcome in *Azotobacter vinelandii* by additions of magnesium ions (JOHNSON and COLMER 1957 and 1958) and has been partially overcome in lupine seedling roots by the addition of ascorbic acid (WEST and HENDERSON 1950).

The uptake of glucose-U-C^{14} has been observed to be three times greater in first internode bean stem sections from 2,4-D treated plants than from controls (FANG et al. 1960). The carbon dioxide production from these segments was also enhanced. Following 2,4-D treatment of bean stem segments, FANG et al. (1961) noted an increase in substrate absorption, an increase in respiratory carbon dioxide production, and a decrease in the incorporation of acetate carbon into cellular constituents. They also noted that the formation of lipids by multiple β-condensation of acetate was not affected whereas other pathways of lipid synthesis were increased.

A two-to-three fold increase in respiratory carbon dioxide production from acetate-1-C^{14}, and about a six-fold increase from acetate-2-C^{14} have been demonstrated by HUMPHREYS and DUGGER (1959) in 2,4-D treated corn roots. Employing pea root tissues STEVENS et al. (1962) showed that pretreatment with 2,4-D caused a greater increase in respiratory carbon dioxide production from acetate-2-C^{14} than from acetate-1-C^{14}. Both biologically active and inactive phenoxy acids inhibited acetate absorption alike and similarly affected acetate utilization causing STEVENS et al. (1962) to conclude that the toxic effect produced by 2,4-D and its active analogs is not due to a disturbance in acetate metabolism.

BOURKE et al. (1962) demonstrated that both herbicidal and non-herbicidal phenoxy acids inhibited glucose absorption by pea roots as well as interfering with glucose metabolism. Low concentrations of 2,4-D stimulated glycolysis while high concentrations inhibited both glycolysis and pentose phosphate cycle activity. Glycolysis apparently was inhibited to a greater extent than the pentose phosphate cycle. The high 2,4-D concentration reduced carbon dioxide output and the assimilation of cell wall material.

HUMPHREYS and DUGGER (1957 a) found that 2,4-D caused an increase in the catabolism of labeled glucose via the pentose phosphate pathway in root tips of corn, pea, and oat seedlings. In corn, oats, and peas which had their cotyledons removed, glycolysis was unaffected. When the pea cotyledons remained attached, glycolysis was reduced. They further noted that the influence of 2,4-D on glucose metabolism differed from that of IAA.

Following the removal of the cotyledons from etiolated pea seedlings, respiration has been shown to decline rapidly (HUMPHREY and DUGGER

1957 b). However, if the seedlings had been treated with 2,4-D the decline was less rapid. The respiratory quotient values they obtained, indicated that both treated and untreated seedlings respired carbohydrates predominantly. They concluded that the 2,4-D treatment shifted glucose metabolism from the glycolytic pathway to the pentose phosphate pathway (HUMPHREYS and DUGGER 1957 and 1961).

HUMPHREYS and DUGGER (1959) showed that dinitrophenol at 4.5 x $10^{-5}M$ inhibited carbon dioxide production from glucose, pyruvate, and succinate in excised corn roots. If the root tips had received a 2,4-D treatment this inhibition was not evident. The increased respiration by the corn root tips following a 2,4-D treatment was inhibited only a small degree by anti-mycin A (HUMPHREYS and DUGGER 1961). The shift from the glycolytic pathway to the pentose phosphate pathway in corn root following 2,4-D treatment has been postulated by HUMPHREYS and DUGGER (1961) as possibly being due to the presence of an alternate pathway for electron transport to oxygen. In the 2,4-D treated roots a non-mitochondrial electron transport pathway appeared to be present. If indeed this pathway served as an electron acceptor for $NADPH_2$, it could account for the increase in the pentose phosphate pathway activity after 2,4-D treatment. However, the mitochondria may still be the main site of electron transport to oxygen in 2,4-D treated tissues. They assumed that 2,4-D did not affect qualitative changes in the mitochondrial electron transport system.

The results of TOMIZAWA (1956), BLACK (1960), and BLACK and HUMPHREYS (1962) showed decreased activity of the enzymes in the glycolytic pathway following 2,4-D treatment. BLACK (1960) and BLACK and HUMPHREYS (1962) also observed enhanced activity of the enzymes in the pentose phosphate pathway following 2,4-D treatment. BLACK (1960) observed a 2,4-D-induced increased utilization of ribose-5-phosphate, increased formation of heptulose and hexose from ribose-5-phosphate, and an increased rate of oxidation of both glucose-6-phosphate and 6-phosphogluconate in corn seedling tissues. The decreased activity in the glycolytic pathway was evident from the decreased activity of 6-phosphofructokinase, aldolase, and glyceraldehyde-3-phosphate dehydrogenase in extracts from 2,4-D treated corn seedlings. The activity of phosphoglucoisomerase, phosphoglyceric kinase, and enolase were not affected by 2,4-D treatment (BLACK 1960, BLACK and HUMPHREYS 1962).

In culture solution experiments with vetch internodes, SIVORI (1953) found that the following substances inhibited the growth-promoting action of 2,4-D: sodium sulfite, dinitrophenol, potassium cyanide, sodium pyrophosphate, malonic acid, thiourea, sodium fluoride, chromic acid, hydroxylamine, diethyldithiocarbamate, formaldehyde, phenylmercuric acetate, sodium selenite, and sodium salicylate. This would indicate that a certain metabolic integrity is prerequisite for the expression of the 2,4-D induced stimulation. 2,4-D stimulated the respiration in corn coleoptile segments when it promoted growth in length of the segments (FRENCH and BEEVERS 1953).

They assumed these results were caused by an increased supply of high energy phosphate acceptors. BRODY (1952) had found earlier that 2,4-D was an uncoupling agent for oxidative phosphorylation in rat-liver mitochondria. In a phosphate deficient media, the 2,4-D stimulated the respiration of the rat-liver mitochondria. Studying respiratory metabolism in the red kidney bean, BURLINGHAM (1960) concluded that the supply of ATP within the mitochondria was apparently reduced in all instances when 2,4-D was present. Presumably 2,4-D affected the distribution of cellular metabolic energy.

2,4-D did not markedly affect the respiration of isolated mitochondria from the leaves of young pea plants when succinate used as the substrate (KHUBUTIYA 1959). However, KHUBUTIYA (1959) found that oxidative phosphorylation was completely blocked in the presence of $10^{-3}M$ 2,4-D and partially blocked in the presence of $10^{-4}M$ 2,4-D. Similarly STENLID and SADDIK (1962) found that 2 x $10^{-3}M$ 2,4-D caused 50 percent uncoupling of oxidative phosphorylation using succinate as the substrate. The D and L forms of 2,4-DP acted similarly although the first is an auxin and the second is an auxin antagonist. 2,4-D and all the other halogenated phenoxyacetic acid derivatives examined by CHKANIKOV et al. (1964) uncoupled oxidative phosphorylation in isolated mitochondria from soybean hypocotyls, regardless whether or not the compound showed auxin-like activity. The toxicity of the compounds increased with the number of substituents and did not depend on their position. Succinate was again the substrate employed for the isolated mitochondria.

Respiration studies by WEDDING and BLACK (1961) showed that at 2,4-D concentrations which inhibit P^{32} incorporation into nucleotides, the oxygen uptake by Chlorella cells was stimulated, indicating that 2,4-D uncoupled phosphorylation in a manner similar to dinitrophenol. In their experiments, $10^{-4}M$ 2,4-D in the growth medium reduced P^{32} incorporation into ATP and ADP to 30 percent; and that of AMP plus inorganic phosphorus to 85 percent of that in the controls. Total phosphorus was reduced to 76 percent of the controls. Later WEDDING and BLACK (1962) concluded that the severity of uncoupling observed in both monocots and dicots, using a number of substrates for the isolated mitochondria, could account for the phytotoxicity of 2,4-D. Concentrations of 2,4-D capable of causing uncoupling also completely inhibited growth of disks from the same tissues. However, they noted that 2,4-D had only 1/20 the uncoupling activity of 2,4-dinitrophenol.

LOTLIKAR (1960) observed that 2.5 x $10^{-3}M$ 2,4-D completely inhibited oxidative phosphorylation in cabbage mitochondria; however, oxygen uptake was inhibited by only 65 percent. From his studies on oxidative phosphorylation he suggested that 2,4-D interfered with energy transfer at a site proceeding the point of action of dinitrophenol.

Hypocotyls from soybean seedlings sprayed with 5 x $10^{-7}M$ 2,4-D were observed by KEY and GALITZ (1959) KEY (1964 b) to contain an inhibitor of germination which also acted as an uncoupler of mitochondrial oxidative phosphorylation.

SWITZER (1957) found that if $10^{-8}M$ 2,4-D was added to the reaction mixture containing mitochondria isolated from soybeans, the oxygen uptake was reduced by 50 percent and the P:O ratio by approximately 44 percent. Thus, 2,4-D caused *in vitro* uncoupling of oxidative phosphorylation. However, when the plants were treated with 2,4-D before the mitochondria were isolated, there was no observable uncoupling. Instead he observed a small stimulation of oxidative phosphorylation which he attributed to the formative effects of 2,4-D leading to the isolation of particles that retained more of their original activity than those isolated from non-treated plants. KEY et al. (1960) suggested that 2,4-D initiated biochemical syntheses leading to mitochondria which were most efficient in oxidative phosphorylation.

Mitochondria isolated by KEY and HANSEN (1958) from soybean seedlings sprayed with 110 p.p.m. of 2,4-D were found to have higher oxidative and phosphorylative activity than controls. Phosphorus uptake was stimulated more than oxygen uptake. Later investigations (KEY 1959, KEY et al. 1960) showed that mitochondria isolated from the swollen regions of the 2,4-D treated soybean seedling hypocotyl were larger and were more active in oxidative phosphorylation than mitochondria from inhibited apical regions (the hypocotyledonary hook). The increased phosphorylation could be almost completely uncoupled by RNAase. Exogenous AMP was not required for maximum substrate oxidation. They concluded that auxin induced growth involved growth of the mitochondria, since inhibited tissues contained inhibited mitochondria. YAKUSHKINA and LIKHOLAT (1964) found a stimulation of oxidative phosphorylation in mitochondria from sunflower roots. Without a knowledge of the internal concentration of 2,4-D in the treated tissues, comparisons between the effects of 2,4-D on oxidative phosphorylation between monocots and dicots remain equivocal, especially since LIKHOLAT (1964) has shown that growth stimulatory dosages of 2,4-D increased the amount of readily hydrolyzable phosphorus and ATP present. Growth inhibiting concentrations of 2,4-D on the other hand, showed the opposite effects.

XVII. Photosynthesis

Several investigators (PICKETT et al. 1951, TOMISEK et al. 1957) have failed to find any effect of 2,4-D on photosynthesis. However, there is other evidence indicating that chlorophenoxy herbicide applications reduce the rate of photosynthesis.

FREELAND (1948) showed that a 100 p.p.m. 2,4-D application to beans reduced the apparent rate of photosynthesis in two-to-four day experiments. Later experiments (FREELAND 1950) with *Anacharis* showed that 30 p.p.m. of 2,4-D caused a decrease in the rate of photosynthesis. A slow down in the mechanism involved in the biochemical turnover of carbon was suggested by BUTTS and FANG (1956) as perhaps causing the observed 2,4-D induced decrease in the rate of photosynthesis in bean plants. The results of VERNON and ARONOFF (1952) indicated that 2,4-D may affect plants by decreasing the rate of sugar diffusion from the site of photosynthesis into the main translocation stream.

The chlorinated derivatives of some phenoxyacetic acids were examined by MORELAND and HILL (1962) for interference with the HILL reaction of isolated chloroplasts. There appeared to be a correlation between phytotoxicity and the inhibition of the HILL reaction in the chloroplasts.

Applications of one kg./hectare of sodium-2-4-D reduced both photosynthesis and the chlorophyll content in the monocots and dicots studied by MALISAUSKIENE (1959 and 1962). Since the reduction in photosynthesis occurred sooner and was more extreme than the reduction in chlorophyll content, he concluded that reduced photosynthesis was not primarily caused by the reduction in chlorophyll content. LADONIN (1960) measured a 67 percent reduction in chlorophyll content in a number of plants seven days after a foliar 2,4-D application. Photosynthesis was not affected for four days after the herbicide application but dropped markedly after seven days.

Treatment with 2,4-D resulted in a more severe retardation of photosynthesis in oats than in sunflower in the investigation of RAKITIN and POTAPOVA (1959 b). The effect of 2,4-D on photosynthesis in various corn varieties has been used by MASHTAKOV et al. (1962) and MASHTAKOV and PAROMIHIK (1962) as an index of the resistance of the corn variety to 2,4-D injury. A reduction in the rate of photosynthesis appeared to be correlated with the degree of injury in corn. Certain corn varieties showed a greater repression of photosynthesis due to MCPA treatment than to 2,4-D. In contrast to this, MALISAUSKIENE (1961) observed that both a resistant and a susceptible variety of oats showed a reduction in the rate of photosynthesis following the application of one kg./hectare of 2,4-D.

WILLIAMS (1963) observed that carbon dioxide uptake by mustard plants following a treatment of 500 p.p.m. 2,4-D was inhibited to a greater extent in red and pink light than by any other light quality examined. The action of a 100 p.p.m. spray of 2,4-D was not significantly influenced by light quality. The presence of two or five percent sucrose in the 500 p.p.m. 2,4-D spray reduced the effect of the 2,4-D interference with carbon dioxide uptake under all lights. Low light intensity seemed to be more effective than high light intensity in promoting the 2,4-D effects. Following the application of a 1,000 p.p.m. spray of 2,4-D, blue light followed by pink or red light were the most effective in promoting the 2,4-D interference with carbon dioxide uptake. Levels of 2,4-D which stimulated plant growth have been shown to enhance dark fixation of carbon dioxide in bean plants (HUFFAKER et al. 1962).

WEDDING et al. (1954) treated leaves of Washington navel orange and cells of Chlorella pyrenoidosa with 2,4-D. The resulting inhibition of photosynthesis was correlated to the concentration of undissociated 2,4-D acid molecules present in the solution, regardless whether this was obtained by varying the pH of the buffer solutions containing a constant amount of 2,4-D or by varying the concentration of 2,4-D added to the buffers of the same pH. In the presence of low concentrations of 2,4-D in the undissociated acid form, the inhibition of photosynthesis was proportioned to the log of this concentration. At high concentrations of 2,4-D the inhibition was directly proportional to the concentration. WEDDING et al. (1954) suggested that this might be explained as direct or indirect destruction of chlorophyll. ERICKSON et al. (1955) noted that 2,4-D was more than 200 times as effective as acetic acid in inhibiting photosynthesis in *Chlorella*, based on the concentration of undissociated molecules at pH 4.0. But a comparison based on the postulated anion concentration in the cytoplasm indicated that 2,4-D was only three to four times as effective as acetic acid in inhibiting photosynthesis.

XVIII. Relationship of chemical structure to activity

One of the early workers in this field, ZIMMERMAN (1943), showed that the order of increasing activity of phenoxyacetic acids in cell elongation was as follows: phenoxyacetic acid, σ-chlorophenoxyacetic acid, *p*-chlorophenoxyacetic acid, 2,4-D. Bromo-substituted compounds were less active than the chloro-substituted compounds. Subsequently there have been several investigations in this area. VELDSTRA (1953) and WEINTRAUB (1953) have reviewed the literature in regard to structure-activity relationships.

Optimal translocation of 2,4-D takes place under conditions which favor the translocation of photosynthate (LINDER et al. 1949). Translocation does not appear to be greatly influenced by the number of carbons in the ester side chain, nor the amount of growth substance present over a range of ten to 200 μg./leaf. These conditions have facilitated investigations into structure-function relationships; however, the possibility that chlorophenoxy herbicides may exert a multiplicity of actions has made the establishment of relationships between structure and activity more difficult. Certain of the observed physiological actions of herbicides may not be related to auxin activity such as the uncoupling of oxidative phosphorylation by halogenated phenoxyacetic compounds as shown by CHKANIKOV et al. (1964). The uncoupling of oxidative phosphorylation was shown by CHKANIKOV et al. (1964) to increase with the number of substituents and did not depend on their position.

Among the studies on phenoxy herbicides relating structure to activity are those of MCNEW and HOFFMAN (1950). They came to the following conclusions after testing 53 derivatives of 2,4-D on pinto beans and tomatoes. The chlorine in the 2-position did not contribute greatly to the activity and could be replaced by a methyl group, whereas the chlorine in the 4-position was necessary for activity. The phenoxy ring was also necessary for activity and could not be replaced by a methoxy ring without loss of activity. The carboxyl or some other negative group on the side chain appeared to be prerequisite for activity. The hydroxyl group could be replaced in the formation of an amide, salt, or ester without destroying the biological activity.

HANSCH and MUIR (1950) observed that the growth regulator activity of 2,4-D and other similar compounds could be blocked if both *ortho* positions were substituted on the phenoxy ring. They interpreted this as indicating that these plant growth regulators reacted with a plant component or components through an *ortho* position. Further studies (MUIR and HANSCH 1953) involving the examination of the relative activities of 117 compounds on the ability to induce elongation of the *Avena* coleoptile led them to the conclusion that the two-point *ortho* reaction was the best hypothesis. However, some of the compounds they examined failed to conform to this hypothesis. The following rules for auxin or auxin-like activity for synthetic auxins were proposed by THIMANN (1952). First, activity required two free positions *para* to one another and secondly, activity required one free position *ortho* to the carboxyl. The two-point attachment theory has been supported by the experiments of PALEG and MUIR (1952). They failed to find any correlation of surface activity with physiological activity as measured by the suppressive effect of plant growth regulators on the polarograph oxygen maximum. They concluded that chemical reactivity was more important than the ability to be adsorbed for the manifestations of growth-regulating properties by compounds such as 2,4-D and 2,4,5-T. The evidence in favor of the two-point attachment hypothesis as the mode of reactivity has recently been reviewed by HANSCH and MUIR (1961).

A theory on auxin action presented by VAN OVERBEEK (1961 b) postulates that the primary mode of action of compounds such as 2,4-D is physico-chemical rather than a chemical nature. The theory discounts the involvement of the *ortho* position or the carboxyl group in a covalent bond with the substrate. Instead it promotes the idea that auxin molecules can solubilize in the cytoskeleton. Since the membrane of the cytoskeleton is probably lipoprotein in nature, the auxin and growth regulator would be required to have a partitioning coefficient favorable for partitioning into fats. The various constituent groups of the molecule required for auxin activity were explained as being required for positioning the auxin molecule in the cytoskeleton.

Some physical-chemical aspects of the mode of action of synthetic auxins have been investigated by FREED *et al.* (1961 b). They concluded that the mechanism of 2,4-D action is one of adsorption on a protein surface, as a result of this the protein structure is modified with a consequential change in its enzymatic activity. Enzymes they examined were stimulated in activity following treatment with 2,4-D concentrations of 40 to 100 p.p.m. At concentrations of 500 to 1,000 p.p.m. they were inhibited. They observed decreased 2,4-D inhibition of mitochondrial respiratory activity as the temperature was increased from 25° to 30°C. Thus they concluded that the mechanism of action is a physical adsorption of 2,4-D to various proteins. They inferred that not all enzymes or proteins adsorbed the 2,4-D with equal facility, thus allowing for marked differences in enzyme response to 2,4-D. Similarly the same enzyme derived from different sources might have different affinities for 2,4-D.

It appears there are a number of correlations between molecular structures and auxin activity. For example, COLINESE et al. (1962) observed that in substituted phenoxyacetic acids including 2,4-D, the plant growth regulating activity increased with increasing dipole moment. SUDI et al. (1961) examined various amides of 2,4-D and concluded that the presence of a carboxyl group was prerequisite for auxin activity. Their conclusions were based on the evidence that treatment with the amide was followed by the production of the acid and in amounts approximately proportional to the auxin effects shown by the plants treated.

Employing the theory of linear combinations of atomic orbitals, JULG and COCORDANO (1962) found auxin activity of chlorinated derivatives of phenoxyacetic acid to vary in the same way as the sum of the free valence indices at positions 3 and 6. They postulated that the auxin molecules, perhaps attached by this acid group to a cellular component or components are deposited and attached there by two points, namely positions 3 and 6. The attachment may be through a large atom like the sulphur of cystine present in the cells. COCORDANO and RICARD (1963) applied quantum mechanics to this problem and reached the same conclusions as JULG and COCORDANO (1962). JOHNSTON (1962) related growth-regulating activity of substituted amides of 2,4-D to the electronic status of the C-O-N portion of the molecules. He advanced the theory that an essential structural feature of an active auxin molecule is an electron-donating group or nucleophillic center at a particular position in the side chain of the molecule, not necessarily a carboxyl acid group. Low activity of apparent exceptions to this theory are explained as being due to competing donor groups which lead to misalignment of the molecule on the receptor centers in the plant.

Quite possibly the exact structural requirements for growth-regulating activity will only be determined when the mode of action of these growth regulators has been firmly established.

XIX. Fate of herbicides in plants

The fate of chlorophenoxy herbicides has been reviewed by AUDUS (1961), CRAFTS (1964), FREED and MONTGOMERY (1963), HILTON et al. (1963), SHAW et al. (1960), and most recently by SWANSON (1965). Chlorophenoxy herbicides such as 2,4-D may persist in the treated plants for varying periods of time depending upon the physiological state of the plant and the plant species examined. The free 2,4-D in the tissue is deemed to exert its influence as long as it is present. BACH and FELLIG (1961 a) observed that callus growth of bean stem tissue was stimulated by 2,4-D but this stimulation stopped when there was no more free 2,4-D present in the sections.

The period of time the 2,4-D stimulus was observable in cotton was shown by MCILRATH et al. (1951) to be controlled by the extent of the vegetative growth of the cotton. With little or no vegetative growth the 2,4-D stimulus persisted up to eight weeks. Later MCILRATH and ERGLE (1953) extracted a non-naturally occurring growth regulator from cotton plants 80 days after 2,4-D treatment. This compound induced malformations similar to those caused by 2,4-D; however, they did not prove that the compound they isolated was actually 2,4-D. The meristems of non-treated scions were grafted by MUZIK and WHITWORTH (1963) on to tomato plants treated by dipping one leaf in a solution containing 1,000 p.p.m. of 2,4-D. These non-treated meristems produced malformations when grafted on to treated plants 60 days after treatment, but failed to produce malformations 90 days after treatment. Residual 2,4,5-T has been shown to persist on the leaf surface of Tilton apricots for at least one month following foliar applications of C^{14}-2,4,5-T at the rate of 150 p.p.m. (MAXIE et al. 1962). No evidence was obtained for the metabolism of 2,4,5-T in either the leaves or fruits of apricots.

The interpretation of the different physiological responses of chlorophenoxy herbicide resistant plants versus the susceptible plants must always take into account that the internal concentrations of the herbicide may not be similar in the plants. As has been pointed out in the foregoing discussion, physiological responses of plants vary considerably with the concentration of herbicide present. Low concentration may stimulate or affect the response in one fashion whereas high concentration of herbicide may produce the opposite response. Consequently a differential response obtained between chlorophenoxy herbicide resistant and susceptible plants may be only a reflection of the different internal concentration of the herbicide. BOYLE (1954) showed that 2,4-D application could produce the same responses in oats and beans (Phaseolus vulgaris); however, it required five times the amount of 2,4-D to produce the response in oats. The pattern of herbicide distribution was similar in both plants. ASHTON (1958) and FANG and BUTTS (1954 a) agree that the rate of 2,4-D translocation is slower in monocots than in dicots, and that the concentration of 2,4-D was lower in the tissues of monocots than in dicots following a uniform 2,4-D treatment. Differences in internal 2,4-D concentration between resistant and susceptible plants could also result from a more rapid rate of detoxification in the resistant plants as pointed out by SLIFE et al. (1962). The differences in 2,4-D susceptibility in two strains of bindweed were shown by WHITWORTH (1961) not to be the result of differences in entry of 2,4-D into the plant or translocation of 2,4-D in the stem. ASHTON (1959) has pointed out that the 2,4-D content of any plant or tissue is dependent on the absorption, translocation, and metabolism of the herbicide. The relative rates of these processes determine the level of 2,4-D which in turn is responsible for the observed physiological responses.

In a discussion of the fate or metabolism of chlorophenoxy herbicides one must consider that the interaction of the herbicide with the cellular components may result in both toxication and detoxification reactions.

The hypothesis that there is an interaction between 2,4-D and a protein or amino acid following 2,4-D entry into the plant has gained considerable acceptance. Certain investigators, for example, ANDREAE and GOOD (1957), BUTTS and FANG (1956), and VAN OVERBEEK (1964), suggest that a conjugation of 2,4-D with an amino acid or protein is the means whereby 2,4-D is detoxified. Other investigators, for example, MUIR and HANSCH (1953), HANSCH et al. (1951), and FREED et al. (1961 b), consider the primary mode of action of 2,4-D as involving an interaction of 2,4-D with a protein. This at first appears contradictory but these could occur concurrently.

The observed conjugate may account for the fate of only relatively limited amounts of the herbicide. From their investigation with 2,4-D, ANDREAE and GOOD (1957) found this to be the case and concluded that the persistance of 2,4-D may be one reason for its effectiveness as a herbicide. They treated pea epicotyls for 24 hours with 20 p.p.m. of C^{14} labeled 2,4-D. Following the 2,4-D treatment about 40 percent of the level was found in the fraction containing cell walls and protein. Acetone extraction yielded 2,4-D from this fraction. The filtrate consisted of 95 percent unchanged 2,4-D, the remaining five percent was composed mostly of a 2,4-dichlorophenoxyacetylaspartate-like compound which yielded 2,4-D on hydrolysis. Recently SUDI (1964) found that pretreatment of pea stem tissue with 2,4-D increased from three to five times the amount of indole-3-acetylaspartate formed. The compounds examined including 2,4-D which possessed growth-regulating activity gave this effect whereas those that did not possess growth-regulating activity failed to give this response. He concluded that the main metabolite of IAA or NAA in pea tissue is a compound in which a peptide link is formed between the carboxyl group of the auxin and the amino group of L-aspartic acid. Small amounts of a glucose ester and a aspartic acid complex with 2,4-D have been detected by KLAEMBT (1961). On the other hand, ZENK (1960) was unable to detect measurable amounts of a 2,4-D-glycine conjugate.

KHUBUTIYA (1958) concluded from his studies on the metabolism of 2,4-D in bean plants that 2,4-D appeared to be linked to a plant substance from which it could be freed by hydrolysis. The initial binding of the 2,4-D to protein prior to the metabolism of the herbicide has been suggested by CANNY and MARKUS (1960). The binding of auxins to specific proteins has also been suggested by VAN OVERBEEK et al. (1951). JAWORSKI et al. (1955) failed to detect a difference in the rate of reaction between 2,4-D and plant substrate (formation of a conjugate) in etiolated and normal bean plants.

VAN OVERBEEK (1961 b) followed the line of reasoning that the mode of action of growth regulators was of a physio-chemical nature rather than of a chemical one. At that time he suggested that the auxin molecules might solubilize in the cytoskeleton.

One of the earliest concepts of the mode of action of auxin and 2,4-D is that of the two-point attachment theory of MUIR and HANSCH (1953). They stated that the reaction of the growth regulator with the protein substrate is chemical in nature and can best be explaind by the two-point *ortho* reaction mechanism. They regarded cysteine or a cysteinyl unit of a protein as the most likely site of reaction with the growth regulator (HANSCH *et al.* 1951).

GALSTON and KAUR (1961) found that the effect of 2,4-D had on the heat coagulability of the cytoplasmic protein appeared to be correlated with auxin-induced growth. In their experiments with pea epicotyl sections treated with $3 \times 10^{-5}M$ 2,4-D, they failed to find any complexing of the 2,4-D with cytoplasmic protein. Since then GALSTON *et al.* (1964) have shown that IAA exhibits both *in vivo* and *in vitro* binding to macromolecules of etiolated pea homogenate, the acceptor molecule probably being a protein. Complex formation appears to be composed of two reactions. The conversion of IAA to a derivative of unknown structure by a protein fraction of the pea homogenate, and the attachments of the derivative to another macromolecular fraction resembling RNA. Purified pea RNA complexed with the IAA derivative, to a lesser extent with IAA itself, and with 2,4-D at pH value lower than 4.8. Subsequently it was stable at higher pH values, although the complex was not formed by pH 7.0.

There is also evidence to show that the complexing of 2,4-D with cell constituents such as proteins may be a method of detoxification. BRIAN (1958) found that chlorophenoxy herbicides could be adsorbed to cell constituents at locations other than the essential site of action. He observed a broad correlation between specific resistance to MCPA and the ability of tissue components, presumably protein and lipoproteins, to adsorb the chemical at physiologically inactive sites.

BACH and FELLIG (1961 c) discounted either a glycoside or a peptide as conjugating with C^{14}-2,4-D. They isolated an unknown labeled biologically inactive compound from bean stem sections. This chromatographic spot did not coincide with the carbohydrate and ninhydrin positive material. The complex they obtained yielded 2,4-D on acid hydrolysis. Since it was biologically inactive it appeared to be a detoxification product

BUTTS and FANG (1956) and FANG (1958) have also studied the formation of non-toxic 2,4-D protein complexes in monocots and dicots. The rate of complex formation was more rapid in resistant plants than in susceptible ones. However, there was no correlation between growth regulator toxicity and the rate of complex formation, as non-toxic compounds also rapidly formed protein complexes. One 2,4-D-protein complex was isolated from 2,4-D treated beans and two complexes were isolated from treated tomatoes, corn, and wheat. These complexes contained relatively similar amounts of at least 12 amino acids. A 2,4-D complex isolated by MORGAN and HALL (1963) from cotton was considered to be the same as FANG's unknown one from beans and a 2,4-D complex isolated from sorghum was considered similar to FANG's unknown three.

ZEMSKAYA and RAKITIN (1964) vacuum-infiltrated 2,4-D into sunflower and oat plants. They concluded detoxification of the 2,4-D in the oat plant occurred by a combination or conjugation with protein and by chemical conversion of the 2,4-D molecule. In the sunflower they had evidence for only a chemical conversion or degradation of the 2,4-D molecule. Consequently they concluded these differences accounted for the divergent responses to 2,4-D treatment by monocots and dicots.

VAN OVERBEEK in his 1964 survey of the mechanisms of herbicide action stated that the inactivation of 2,4-D may very well be by complexing or conjugation with proteins. Plants responding in this manner would be tolerant to 2,4-D. The mechanism of action he postulated was that 2,4-D prevented immature cytoplasm from becoming mature. He hypothesized that the extra growth induced by 2,4-D was caused by the increased production of RNA. He ascribed the abnormal quality of this extra growth to a hormonal imbalance resulting from the influence of saturating concentration of 2,4-D. He concluded that a kinin-auxin imbalance might cause the observed effects from added 2,4-D.

FITES et al. (1964) concluded that jimsonweed is an example of detoxification through translocation and excretion. Most of the labeled 2,4-D foliarly applied apparently remained unbound or unmetabolized. In the stem, where tissue proliferation was occurring, there was an accumulation of the radioactivity with about ten percent of the labeled fraction not being free 2,4-D. The net result was a translocation of 2,4-D from the treated leaf into the root medium. When the 2,4-D concentration in the apex had fallen sufficiently low, apical growth resumed.

The possibility that a herbicidally inactive chlorophenoxy compound could be metabolized or degraded to give rise to an herbicidally active compound was suggested by SYNERHOLM and ZIMMERMAN (1947). They suggested that the omega phenoxybutyric acid might be degraded in plant tissues by β-oxidation, resulting in the conversion of the inactive butyric derivative to the active acetic form. This conversion has also been observed by FAWCETT et al. (1954) and CONWAY (1956). This early work stimulated a number of investigations, these have been reviewed by WAIN (1961) and LINSCOTT (1964). From earlier investigations (FAWCETT et al. 1954, WAIN 1955) WAIN (1961) pointed out that a plant's ability or lack of ability to degrade a phenoxy carboxylic acid such as 2,4-DB via β-oxidation was a possible basis for herbicidal selectivity. The nature and the position of the substituents of the phenoxy compound may well influence the plant's ability to degrade the side chain. An acetone extract was prepared from peas by FREED et al. (1961 a) which had the enzymatic capability of degrading 2,4-DB via β-oxidation. WAIN (1955) suggested from his experiments that MCPB could be converted to MCPA. This has been confirmed in snap beans by BACHE et al. (1964).

Plants apparently are also capable of hydrolyzing the long chain esters of 2,4-D. Castor bean lipase was shown by HAGEN et al. (1949) to hydrolyze the 2,4-D butyl ester to 2,4-D. MORRÉ and ROGERS (1960) demonstrated the hydrolysis of the propylene glycol butyl ether ester and the octyl ester of 2,4-D; similarly SZABO (1963) has observed the hydrolysis of butoxy-ethanol and propylene glycol butyl ether esters of 2,4-D to the acid. This hydrolysis appeared to occur on the leaf surface as well as within the corn and bean plants.

Since the early studies of HOLLY et al. (1950) which suggested that 2,4-D was decarboxylated by red kidney bean plants as a means of 2,4-D degradation there has been considerable conjecture as to the significance of the decarboxylation. WEINTRAUB et al. (1952 a) showed that $C^{14}O_2$ was evolved by bean plants treated with 2,4-D labeled in either the carboxyl or the methylene position. The evolution of $C^{14}O_2$ continued at a relatively low rate for several days after treatment. The initial rate of $C^{14}O_2$ evolution from the carboxyl labeled 2,4-D was several times as great as that from the methylene labeled 2,4-D. $C^{14}O_2$ was not evolved from ring labeled 2,4-D-1-C^{14}. In their investigations, BACH and FELLIG (1961 a, 1961 c) considered the decarboxylation of 2,4-D and the release of $C^{14}O_2$ from carboxyl labeled 2,4-D only a minor pathway of 2,4-D degradation.

In *Quercus marilandica*, BASLER (1964) found only one percent of the carboxyl labeled 2,4-D-C^{14} and only trace amounts of carboxyl labeled 2,4,5-T-C^{14} were decarboxylated. The rates of decomposition of carboxyl labeled C^{14}-2-4-D did not appear to be related to its phytotoxicity in excised leaves of woody plants. Pretreatments of red kidney beans with a 100 p.p.m. aqueous solution of the potassium salt of gibberellin did not affect the rate of degradation of 2,4-D labeled with C^{14} in the carboxyl group according to ASHTON (1959).

CANNY and MARKUS (1960) applied carboxyl labeled 2,4-D to the leaflets of tick beans and determined the rates of degradation of 2,4-D as $C^{14}O_2$ from the various plant parts. Their data indicated that most of the observed 2,4-D breakdown was occurring in the roots of intact plants. Total loss of C^{14} in four days corresponded to about five percent of the applied dose, apparently the main pathway of 2,4-D inactivation in the roots was not by loss of the side chain. When they used tissue sections no differences were measureable between shoots and roots in their capabilities to degrade 2,4-D. GAGARINA (1964) noted a rapid decomposition of 2,4-D in corn but a slow decomposition of 2,4-D in beans. Not only may there be a difference between plant organs in their ability to decarboxylate phenoxyacetic acids, but the substituent groups on the ring of the phenoxyacetic acid may also influence the rate of decarboxylation. Carboxyl labeled 2-chloro-4-fluorophenoxyacetic acid and 2,4-D were applied to the leaves of McIntosh and Stayman apple trees (EDGERTON and HOFFMAN 1961). In 24 hours only four percent of the C^{14} was recovered as $C^{14}O_2$ from the McIntosh variety treated with 2-chloro-4-fluorophenoxyacetic acid compared with a 33 percent recovery from those treated with 2,4-D-C^{14}. Furthermore, EDGERTON (1961) has shown that the resistant McIntosh variety decarboxylated 2,4-D more rapidly than the susceptible Stayman variety. In contrast MORGAN and HALL (1963) found that 2,4-D susceptible cotton decarboxylated 2,4-D-C^{14} five to ten times faster than 2,4-D resistant sorghum. The evolution of $C^{14}O_2$ from wild cucumber *Sicyos augulatus* has been shown by SLIFE et al. (1962) to be approximately ten times greater than the carboxyl labeled 2,4-D than from carboxyl labeled 2,4,5-T. They concluded that the greater phytotoxicity of 2,4,5-T seemed to be related to the inability of the plant to detoxify 2,4,5-T as rapidly as 2,4-D.

The suggestion that the acetate radical of 2,4-D is removed intact and subsequently metabolized are based on evidence such as has been presented by MORGAN and HALL (1963). They showed that the carboxyl label from 2,4-D-C^{14} was more readily decarboxylated than the methylene label. Employing C^{14} and Cl^{36} LEAFE (1962) showed that the resistance of *Galium aparine* to MCPA is due to detoxification of the molecule involving the loss of the 2-carbon side chain rather than by direct decarboxylation. However, MCPP was not metabolized and consequently *G. aparine* was susceptible to this compound. Apparently, the lengthening of the alkyl group by a single methylene group blocked the metabolism.

Red current is resistant to 2,4-D damage, but the black current is not. LUCKWILL and LLOYD-JONES (1960 a) associate toxic action of 2,4-D with the presence of free 2,4-D in the tissues. They noted that the red current leaf could oxidize approximately 50 percent of the carboxyl and 20 percent of the methylene carbon from the 2,4-D side chain over the period of one week. The black current could breakdown only two percent of the 2,4-D in a similar fashion. They observed that of the 2,4-D absorbed by the leaves, five to ten percent of the 2,4-D was detoxified by conversion to water soluble compounds which yielded 2,4-D upon hydrolysis with 2N sulphuric acid. Of the absorbed 2,4-D, ten to 30 percent was bound into the leaf tissue and could not be removed by aqueous or organic solvents or by mild hydrolysis. In another experiment LUCKWILL and LLOYD-JONES (1960 b) investigated the ability of a resistant and a susceptible variety of apple to decarboxylate 2,4-D. The resistant variety, Cox, could decarboxylate 57 percent of the 2,4-D in 92 hours in the excised leaf experiment, whereas the susceptible variety, Bramley Seedling, could only decarboxylate two percent of the 2,4-D in this period of time. 2,4-D resistance in strawberries and *Syringa vulgaris* has also been shown to be associated with the ability of the resistant plants to decarboxylate 2,4-D (LUCKWILL and LLOYD-JONES 1960 b). EDGERTON (1961), also working with apples, showed that the resistant variety, McIntosh, decarboxylated 2,4-D more rapidly than did susceptible varieties. Sixteen other species were examined by LUCKWILL and LLOYD-JONES (1960 b) but all showed slow rates of decarboxylation regardless whether they were 2,4-D resistant or not.

The formation of metabolites from 2,4-D was suggested as early as 1950 by HOLLEY et al. They noted that seven days after treatment more than half of the recovered C^{14}-labeled material was not C^{14}-2,4-D. WEINTRAUB et al.

(1950) were able to recover approximately 45 percent of the 2,4-D they applied to the primary leaves of bean seedlings 24 hours after herbicide treatment. Later WEINTRAUB et al. (1952 b) found a portion of the C^{14} to be present as a relatively volatile or unstable ethanol-soluble acidic material. AUDUS and THRESH (1956 a and b) suggested that a certain amount of the absorbed 2,4-D had been converted to a neutral detoxification product in the plant, this product decomposed to liberate 2,4-D during the chromatographic analysis. At temperatures between 70° and 85° F. MORTON and MEYER (1962) found 80 percent of the labeled 2,4,5-T was metabolized to unidentified products in 25 hours.

The degradation complex of 2,4-D in cotton was demonstrated by MORGAN and HALL (1963) to be different than the one in sorghum. At least three major unidentified metabolites of C^{14}-2,4-D and C^{14}-2,4,5-T have been obtained by BASLER et al. (1964). SLIFE et al. (1962) working with wild cucumber observed that about 75 percent of the absorbed 2,4-D was converted to two major metabolites within 24 hours, but that eight days after 2,4-D treatment there still remained measurable quantities of 2,4-D in the plant tissues. JAWORSKI and BUTTS (1952) found three unknown metabolites of 2,4-D in 80 percent alcohol extracts of bean stems from plants treated with 2,4-D labeled in the carboxyl or methylene position. On the basis of their chromatographic results they suggested that one of these metabolites was a glycoside with 2,4-D as the aglycon. This metabolite was formed in relatively large amounts and was quite stable.

One week after treatment of beans with labeled 2,4-D, HOLLEY (1952) found that a water soluble radioactive material made up 60 percent of the radio-activity present. This material could be detected as soon as six hours after 2,4-D treatment. It was hydrolyzed by acid or alkali to an ether extractable organic acid which differed from 2,4-D. HOLLEY (1952) thought it might be 3-, 5-, or 6-hydroxy-2,4-dichlorophenoxyacetic acid.

Stem tissue from four-week old bean plants was treated with carboxyl labeled C^{14}-2,4-D by BACH (1961) and BACH and FELLIG (1961 b). Two days later, half of the label was found in ten components of an ether extract. The components retained the aromatic nucleus of the 2,4-D. Neither 2,4-D nor 6-hydroxy 2,4-D were isolated. The ether insoluble fraction yielded at least six major components.

Recently CROSBY (1964) has demonstrated by microcoulometic gas chromatography that the controversial ether-soluble substances in plant extracts are conclusively 2,4-D itself. The ether-insoluble fraction yielded 2,4-D and another chlorinated compound on acid hydrolysis, but neither 2,4-dichlorophenol nor its derivatives could be detected.

Analyzing infrared spectra, THOMAS et al. (1965 b) determined that phenoxyacetic acids with unsubstituted 4-positions were hydroxylated at this point by oat mesocotyl tissue and stored as the 4-O-β-D glucoside. They showed this for 2-chloro- and 2,6-dichloro- derivatives of phenoxyacetic acids. 2,4-D and 4-CPA, which are substituted in the 4-position, were not hydroxylated but were converted to phenoxyacetylglucoses. In stem tissue of *Phaseolus vulgaris*, THOMAS *et al.* (1964 a) showed that 2,4-D was converted to a mixture of 2,5-dichloro-4-hydroxyphenoxyacetic acid and 2,3-dichloro-4-hydroxyphenoxyacetic acid. These products accumulated as the β-glucosides. Prior to this, WILCOX *et al.* (1963) had obtained evidence that indicated phenoxyacetic acids could be converted to a metabolite that contained a hydroxyl group in the ring, perhaps 4-hydroxyphenoxyacetic acid by the excised roots of oats, barley, and corn. However, they did not detect any hydroxylated derivatives when the excised roots were supplied with 4-CPA, 2,4-D, or 2,4,5-T.

ZENK (1960) suggested that 2,4-D degradation might proceed by an oxidative pathway. He reported that 2,4-D can enter into reactions resulting in the formation of thioesters of CoA.

Investigating the fate of 2,4-D in bean seedlings, HAY and THIMANN (1956) found that freely extractable 2,4-D rapidly disappeared in the plant, presumably by breakdown. The disappearance of 2,4-D was faster in the light than in the dark, but even in the dark 3/4 of the absorbed 2,4-D disappeared in five days. The inactivation of 2,4-D by light in the presence of riboflavin has been demonstrated by CARROLL (1952), BELL (1956), and HANSEN and BUCKHOLTZ (1952). Phenols have been suggested by HANSEN and BUCKHOLTZ (1952) and shown by BELL (1956) to be the breakdown products following irradiation.

XX. Conclusions

A number of trends are apparent from a survey of the literature concerning the changes in chemical composition of plants following chlorophenoxy herbicide treatment. Among them, observed increases in a number of materials or activities such as phosphorus content, vitamin content, protein content, amino acid content, and proteolytic enzyme activity appear in the plant stem and often in the roots, whereas lowered amounts or activities are apparent in the leaves.

There are several plausible explanations for observed changes in chemical composition. CANNY and MARKUS (1960) have shown that all plant organs do not physiologically respond in a similar manner to 2,4-D treatment. Their experiments appear to indicate that intact plants may respond differently than do tissue sections.

The internal levels of 2,4-D are seldom analyzed when metabolic changes in various tissues or plant parts have been measured. It has been clearly shown by FREED *et al.* (1961 b) and many others that low levels of the chlorophenoxy herbicides may be stimulatory in nature, whereas high levels are inhibitory. Consequently, an observed response in a plant organ is very dependent upon the internal herbicide concentration in that particular organ. Observed differences in physiological responses between resistant and susceptible species may also be due to differences in internal concentration of herbicide. When these differences in internal herbicide concentration are removed, both species may well give similar physiological responses as shown by BOYLE (1954).

Stimulatory levels of chlorophenoxy herbicides appear to cause a reversion of the plant tissues to a meristematic state; this is particularly evident in the stem (CHRISPEELS and HANSON 1962). The herbicide treated area may act like a metabolic sink (OSBORNE and HALLAWAY 1960 and 1961). Furthermore, the action of 2,4-D may give rise to renewed or stimulated metabolic activity with the accumulation of carbon and nitrogenous compounds in the localized 2,4-D treated area at the expense of the non-treated areas (OSBORNE and HALLAWAY, 1960 and 1961). It appears that as 2,4-D is translocated from the leaves via the phloem to the stem and the roots the concentration in the stem vascular parenchyma reaches a stimulatory level which causes the vascular parenchyma to become meristematic. This meristematic activity and subsequent tissue proliferation then creates a metabolic sink in the stem which probably accounts for the altered protein, amino acid, phosphorus, and vitamin content in the stem at the expense of the leaves.

If cell proliferation in vascular tissues was great enough to crush the phloem, then interference in translocation could also cause the observed changes in the levels of various constituents in plant leaves. Indeed, MACLEOD (1964) found that excised leaves underwent the same changes in constituent levels that occurred in attached MCPA-treated leaves.

Bean plants uprooted and left to die on the green house bench were shown by MUZIK and LAWRENCE (1959) to undergo the same changes in root protein and amino acid content as their 2,4-D treated plants. It is quite possible that a number of the measured effects of 2,4-D treatments are similar to those produced by the onset of death in the plant.

Chlorophenoxy herbicide-induced responses may vary with the time elapsed between the herbicide application and the measurement of the responses. In some instances an initial increase in the responses may be followed by a decline below the levels of controls some days later. Consequently measurements made at only one set time following the herbicide treatment would give an incomplete picture. The diversity in the time elapsed before the analyses have been made following chlorophenoxy herbicide applications may also explain part of the variation in observed results.

Data concerning the changes in chemical composition following herbicide application in field plots may be affected by the enhanced crop growth resulting from the removal of weed competitors and consequently be difficult to evaluate.

The large number of physiological activities in the plant affected by chlorophenoxy herbicides indicates that the mode of action of these compounds must either be very non-specific, like the physical-chemical theories of FREED et al. (1961 b) or VAN OVERBEEK (1961 b) or affect a specific step basic to all the observed responses. Such a step would necessarily be one which could control plant metabolism and would very likely involve nucleic acid metabolism. The exact influence of the chlorophenoxy herbicide on the physical and chemical structure of DNA and the consequent quantitative and qualitative alteration in RNA metabolism remains to be determined.

The rate of penetration and translocation of chlorophenoxy herbicides may play a role in the determination of selectivity in certain plant species. The ability of certain resistant plants to detoxify the chlorophenoxy herbicides by degradation of the herbicide molecule or by conjugation of the herbicide with cellular components has also received considerable attention as a partial basis for herbicidal selectivity. The degree of response to the chlorophenoxy herbicides has been correlated with the concentration of the free herbicide present in the plant cell. Therefore, differences in the factors affecting this concentration, such as penetration, translocation, or detoxification, could account for herbicidal selectivity.

Summary

The introduction of the chlorophenoxy herbicides for vegetation management pioneered the modern approach for weed control which has resulted in the establishment of a new scientific discipline, Weed Science. The basic research conducted on the chlorophenoxy herbicides is the foundation of this science and the approaches developed during these investigations have been used numerous times in the research on the many other classes of compounds which have been subsequently developed as herbicides.

The effects these compounds caused were entirely different from anything that plant scientists had encountered previously. Low concentrations had pronounced morphological effects on organs and tissues some distance from the point of application—a "hormone." Their molecular configurations were quite similar to the naturally occurring plant growth substance-indoleacetic acid. In fact they are the progeny of plant growth substance research. They had an applied agricultural use. All of these factors contributed to the great interest of the agronomists, anatomist, physiologist, and biochemist. Therefore, a large number of plant scientists conducted experiments to elucidate the mode of action of these interesting compounds or used them as a new tool to study their particular systems. Consequently, there are a multitude of papers describing their effects. Many of these including biochemical and metabolic changes are in the area covered by this review.

Although this review includes all of the chlorophenoxy herbicides, 2,4-D was the most commonly used compound. It should also be noted that there is still a considerable amount of research being conducted on the mode of action of the chlorophenoxy herbicides and although several mechanisms have been proposed none of these have been universally accepted.

This review discusses the changes in composition of carbohydrates, lipids, nitrogen, organic acids, ethylene, alkaloids, steriods, aromatics, vitamins, pigments, minerals, water, auxins, nucleic acids and enzymes induced by the chlorophenoxy herbicides. It also discusses the metabolic changes observed in nitrogen metabolism respiration, photosynthesis, and nucleic acid metabolism, respiration, photosynthesis, and nucleic acid metabolism due to these herbicides. Brief consideration has been given to fate of these herbicides as well as to the relationship of the biological activity to their chemical structure.

Résumé*

L'introduction des herbicides à groupe chlorophénoxy pour le traitement de la végétation a constitué la première matérialisation de la méthode moderne de lutte contre les mauvaises herbes ayant abouti à l'apparition d'une nouvelle discipline scientifique, la Science des plantes nuisibles. Les recherches fondamentales affectuées avec les herbicides à groupe chlorophénoxy constituent la base de cette science et les directions explorées lors de ces investiagtions ont été, à maintes reprises, appliquées à la recherche concernant les nombreuses autres classes de composés qui ont ensuite été proposés comme herbicides.

Les effets provoqués par ces composés se sont révelés tout à fait différents de ceux que les spécialistes des problèmes végétaux étaient habitués à observer. De faibles concentrations exercent des effets morphologiques importants sur des organes et des tissus éloignés du point d'application, comparables à ceux d'une "hormone." Leurs structures moléculaires ressemblent beaucoup à celle de l'acide indolacétique excitant de la croissance des végétaux existant à l'état naturel. En fait, ils constituent une lignée pour la recherche sur les substances de croissance chez les végétaux. Ils ont connu des applications en agriculture et ont éveillé un grand intérêt chez les agronomes, les anatomistes, les physiologistes et les biochemistes. Par suite, de nombreux spécialistes de la science des végétaux ont effectué des expériences tendant à élucider le mode d'action de ces intéressants composés et les ont utilisé comme de nouveaux outils pour étudier leurs systèmes particuliers. Il en résulte une multitude de publications décrivant leurs effets. Beaucoup d'entre eux comprenant les modifications biochimiques et métaboliques s'insèrent dans le cadre du sujet traité dans cette revue. Bien qu'elle traite de tous les herbicides à groupe chlorophénoxy, le 2.4 D constitue le composé le plus couramment utilisé. Il faut noter qu'il y a encore use masse de recherches en cours sur le mode d'action des herbicides à groupe chlorophénoxy et, bien que plusieurs mécanismes aient été proposés, aucun d'entre eux n'a été universellement accepté.

Dans cette revue sont discutées les modifications dans la composition en glucides, lipides, produits azotés, acides organiques, dérivés éthyléniques, alcaloïdes, stéroïdes, produits odorants, vitamines, pigments, constituants minéraux, eau, auxines, acides nucléïques et enzymes, provoquées par les herbicides à groupe chlorophènoxy. Sont discutées également les modifications métaboliques observées dans le métabolisme de l'azote, la respiration, la photosynthèse et le métabolisme des acides nucléiques, dûes à ces her-

* Traduit par R. TRUHAUT.

bicides. Il est procédé à un examen sommaire du sort de ces herbicides, ainsi que des relations entre leur activité biologique et leur structure chimique.

Zusammenfassung*

Die Einführung der Chlorphenoxy-Herbizide zur Behandlung des Pflanzenwuchses war bahnbrechend für einen modernen Weg der Unkrautbekämpfung, der auf die Errichtung einer neuen wissenschaftlichen Disziplin, der Unkrautwissenschaft, hinauslief. Die Grundlagenforschung, die zu den Chlorphenoxy-Herbiziden führte, ist das Fundament dieser Wissenschaft, und die während dieser Untersuchungen entwickelten Wege wurden in zahlreichen Fällen in der Erforschung vieler anderer Verbindungsklassen, die später als Herbizide entwickelt wurden, verwendet.

Die von diesen Verbindungen verursachten Effekte waren ganz verschieden von allem, was dem Pflanzenwissenschaftler vorher begegnet war. Niedrige Konzentrationen hatten ausgeprägte morphologische Effekte auf Organe und Gewebe, in gewisser Entfernung vom Ort der Anwendung—wie bei einem "Hormon." Ihre Molekülkonfiguration war ganz ähnlich dem natürlich vorkommenden Pflanzenwirkstoff Indolessigsäure. Tatsächlich waren sie die Nachkommen der Pflanzenwirkstoff-Forschung. Sie hatten eine gebräuchliche Anwendung in der Landwirtschaft. Alle diese Faktoren steuerten zum grossen Interesse von seiten der Agronomen, Anatomen, Physiologen und Biochemiker bei. Darum führte eine grosse Zahl von Pflanzenwissenschaftlern Untersuchungen durch, um die Aufklärung der Wirkungsweise dieser interessanten Verbindungen zu versuchen oder sie benutzten sie als neues Mittel zum Studium ihrer besonderen Systeme. Folglich gibt es eine Vielzahl von Abhandlungen, die ihre Effekte beschreiben. Viele davon, einschliesslich der Abhandlungen über Veränderungen der Biochemie und des Stoffwechsels werden in dieser Übersicht behandelt.

Obgleich diese Übersicht alle Chlorphenoxy-Herbizide einschliesst, war 2,4-D die üblicherweise am meisten verwendete Verbindung. Es wäre auch zu bemerken, dass noch ein bedeutender Betrag an Forschung über die Wirkungsweise der Chlorphenoxy-Herbizide durchgeführt werden muss und dass trotz des Vorschlages verschiedener Mechanismen keiner allgemein angenommen wurde.

Diese Übersicht diskutiert die Veränderungen des Gehalts an Kohlenhydraten, Lipiden, Stickstoff, organischen Säuren, Alkaloiden, Steroiden, Gewürzen, Vitaminen, Pigmenten, Mineralstoffen, Wasser, Auxinen, Nucleinsäuren and Enzymen, hervorgerufen durch die Chlorphenoxy-Herbizide. Sie diskutiert auch die durch diese Herbizide bedingten Stoffwechseländerungen, unter Beobachtung von Stickstoff-Stoffwechsel, Atmung, Photosynthese und Nucleinsäurestoffwechsel. Eine kurze Berücksichtigung fand das Schicksal dieser Herbizide, sowie die Beziehung zwischen biologischer Wirksamkeit und ihrer chemischen Struktur.

---

* Übersetzt von F. BÄR.

References

AKERS, T. J., and S. C. FANG: Plant metabolism. VI. Effect of 2,4-D on the metab-
olisom of aspartic acid and glutamic acid in the bean plant. Plant Physiol. **31**,
34 (1956).

ALABUSHEV, V. A.: Effect of the herbicide 2,4-D on the protein N content in barley,
corn, and millet grain. Fiziol. Rast. **9**, 372 (1962).

ALBERSHEIM, P.: Auxin-induced product inhibition of pectin transeliminase as shown
by ozonolysis. Plant Physiol. **38**, 426 (1963).

—, and U. KILLAS: Studies relating to the purification and properties of pectin transeliminase. Arch. Biochem. Biophys. **97**, 107 (1962).

ALLEN, F. W.: The influence of growth regulator sprays on the growth, respiration,
and ripening of Bartlett pears. Proc. Amer. Soc. Hort. Sci. **62**, 279 (1953).

ANDREAE, W. A., and S. R. ANDREAE: Studies on indoleacetic acid metabolism. I.
Effect of 4-methylumbelliferone, maleic hydrazide, and 2,4-D on indoleacetic acid
oxidation. Can. J. Bot. **31**, 426 (1953).

—, and N. E. GOOD: 3-Indoleacetic acid metabolism. IV. Conjugation with aspartic
acid and ammonia as processes in the metabolism of carboxylic acids. Plant
Physiol. **32**, 566 (1957).

ASHTON, F. M.: Absorption and translocation of radioactive 2,4-D in sugarcane and
bean plants. Weeds **6**, 257 (1958).

— Effect of gibberellic acid on absorption, translocation, and degradation of 2,4-D
in red kidney bean. Weeds **7**, 436 (1959).

AUDUS, L. J.: Metabolism and mode of action. In: Handbuch der Pflanzenphysiologie.
W. RUHLAND, ed. Vol. XIV, pp. 1055-1083. Berlin: Springer-Verlag 1961.

—, and R. THRESH: Effects of synthetic growth-regulator treatments on the levels of
free endogenous growth-substances in plants. Ann. Bot. **20**, 439 (1956 a).

— — Effects of synthetic growth substances on the level of endogenous auxins in
plants. In: Chemistry and mode of action of plant growth substances. Proc. Symposium London, pp. 248-252 (1956 b).

BACH, M. K.: Metabolites of 2,4-dichlorophenoxyacetic acid (2,4-D) from bean
stems. Plant Physiol. **36**, 558 (1961).

—, and J. FELLIG: Correlation between inactivation of 2,4-dichlorophenoxyacetic acid
and cessation of callus growth in bean stem sections. Plant Physiol. **36**, 89
(1961 a).

— — Metabolism of carboxyl-C^{14}-labeled 2,4-dichlorophenoxyacetic acid (2,4-D)
by bean stems: heterogeneity of the ethanol-soluble, ether-insoluble products. Nature **189**, 763 (1961 b).

— — The uptake and fate of C^{14}-labeled 2,4-dichlorophenoxyacetic acid in bean
stem sections. In: Plant growth regulation. Fourth Internat. Conf. on Plant
Growth Regulation, pp. 273-287. Ames: Iowa State Univ. Press, 1961 c.

BACHE, C. A., D. J. LISK, and M. A. LOOS: Electron affinity residue determination
of nitrated MCP [2-methyl-4-chlorophenoxyacetic acid], MCPB [4-(2-methyl-4-
chlorophenoxy) butyric acid], and NAA [α-naphthaleneacetic acid]; conversion of
MCPB to MCP in bean plants. J. Assoc. Official Agr. Chemists **47**, 348 (1964).

BARNES, I. C., A. I. McMULLEN, and J. McDONNELL: Hormone influence on the
thiol content of *Hevea brasiliensis* latex. Nature **194**, 1083 (1962).

BASLER, E.: Decarboxylation of phenoxyacetic acid herbicides by excised leaves of
woody plants. Weeds **12**, 14 (1964).

— Interaction of 2,4-dichlorophenoxyacetic acid, ethylenechlorohydrin, and glutathione in the stability of the microsomal ribonucleic acid of excised cotton cotyledons. Proc. Okla. Acad. Sci. **43**, 35 (1962).

—, and T. L. HANSEN: Effects of 2,4-dichlorophenoxyacetic acid and sucrose on
orotic acid uptake in nucleic acids of excised cotton cotyledons. Bot. Gaz. **125**, 50
(1964).

—, C. C. KING, A. A. BADIEI, and P. W. SANTELMANN: Breakdown of phenoxy herbicides in blackjack oak. Proc. 17th Weed Conf., pp. 351-355 (1964).

—, and K. NAKAZAWA: Effects of 2,4-D on nucleic acids of cotton cotyledon tissue. Bot. Gaz. 122, 228 (1961).

BASS, S. T., C. L. HAMNER, and H. M. SELL: Effects of 2,4-dichlorophenoxyacetic acid on the mineral contents of cranberry bean plants (Phaseolus vulgaris). Mich. State Univ. Agr. Expt. Sta. Quart. Bull. 42, 43 (1959).

BASS-BECKING, L. G. M., and R. G. EVERSON: Tissue tensions in Bryophyllum calycinum. Australian J. Biol. Sci. 6, 347 (1953).

BEAUCHAMP, C. E.: Effects of synthetic hormones on the sucrose content of sugar cane. Sugar 46, 42 (1951).

BEEVERS, L., D. M. PETERSON, J. C. SHANNON, and R. H. HAGEMAN: Comparative effects of 2,4-dichlorophenoxyacetic acid on nitrate metabolism in corn and cucumber. Plant Physiol. 38, 675 (1963).

BELL, G. R.: Photochemical degradation of 2,4-dichlorophenoxyacetic acid and structurally related compounds in the presence and absence of riboflavin. Bot. Gaz. 118, 133 (1956).

BENDANA, F., A. W. GALSTON, R. KAUR-SAWHNEY, and P. J. PENNY: Recovery of labelled RNA following in vivo administration of labelled IAA to green pea stem sections. Plant Physiol. 39, suppl. xxxi (1964).

BEREZOVSKII, M. Y., and V. F. KUROCHKINA: Effect of 2,4-dichlorophenoxyacetic acid on the transformation of phosphorus compounds in plants. Doklady Moskov. Sel'skokhoz. Akad. im. K. A. Timiryazeva 25, 182 (1956).

— — Effect of 2,4-D on transformation of phosphorus compounds in plants. Doklady Akad. Nauk S.S.S.R. 113, 458 (1957).

BERG, R. T., and L. W. McELROY: Effect of 2,4-D on the nitrate content of forage crops and weeds. Can. J. Agr. Sci. 33, 354 (1953).

BHARDWAJ, S. N.: Influence of pre-treating the seeds with synthetic phyto-hormone on yield and growth of wheat. J. Indian Bot. Soc. 41, 326 (1962).

BILLERBACK, F. W., N. W. DESROSIER, and R. B. TUKEY: Influence of 2,4,5-trichlorophenoxypropionic acid on the color of red-fruited apple varieties. Proc. Amer. Soc. Hort. Sci. 61, 175 (1953).

BLACK, C. C., JR.: Some effects of 2,4-dichlorophenoxyacetic acid on the carbohydrate metabolism of etiolated corn seedlings. Ph.D. Dissertation. Univ. of Florida, Gainesville (1960).

—, and T. E. HUMPHREYS: Effects of 2,4-dichlorophenoxyacetic acid on enzymes of glycolysis and the pentose phosphate cycle. Plant Physiol. 37, 66 (1962).

BLACKMAN, G. E.: Interrelations between the uptake of 2,4-dichlorophenoxyacetic acid, growth, and ion absorption. In: Chemistry and mode of action of plant growth substances. Proc. Symposium, Kent, England, pp. 253-259 (1956).

BOURKE, J. B., J. S. BUTTS, and S. C. FANG: Effects of various herbicides on glucose metabolism in root tissue of garden peas, Pisum sativum. I. 2,4-Dichlorophenoxyacetic acid and its analogs. Plant Physiol. 37, 233 (1962).

BOYLE, F. P.: Physiology and chemistry of 2,4-dichlorophenoxyacetic acid action on resistant and non-resistant plants. Congr. internat. bot., Paris. Rapps et communs. 8, sect. 11/12, pp. 184-185 (1954).

BRADBURY, D., and W. B. ENNIS, JR.: Stomatal closure of kidney bean plants treated with amonium 2,4-dichlorophenoxyacetate. Amer. J. Bot. 39, 324 (1952).

BRIAN, R. C.: On the action of plant growth regulators. II. Adsorption of MCPA to plant components. Plant Physiol. 33, 431 (1958).

— The effects of herbicides on biophysical processes in the plant. In: Physiology and biochemistry of herbicides. L. J. Audus, ed. pp. 343-356. New York: Academic Press (1964).

BRODY, T. M.: Effect of certain plant-growth substances on oxidative phosphorylation in rat-liver mitochondria. Proc. Soc. Expt. Biol. Med. 80, 533 (1952).

BROWN, J. W.: Effects of 2,4-dichlorophenoxyacetic acid on the water relations, the accumulation and distribution of solid matter, and the respiration of bean plants. Bot. Gaz. **107**, 332 (1946).

BURLINGHAM, G. K.: Effects of 2,4-dichlorophenoxyacetic acid on the growth and respiratory metabolism of tissue from the red kidney bean plant and of the antagonistic effects of uncoupling agents toward the growth-promoting effects of auxin substance. Ph.D. Dissertation. Cornell Univ., Ithaca, N.Y. (1960).

BURSTROM, H.: Activity of plant growth regulators. Ann. Appl. Biol., Proc. 1954 Jubilee Meeting **42**, 158 (1955).

BUTTS, J. S., and S. C. FANG: Tracer studies on the mechanism of action of hormonal herbicides. U.S. Atomic Energy Comm. TID-7512, pp. 209-214 (1956).

BYNOV, F. A., and O. V. TITOVA: Effect of 2,4-D on catalase activity and growth of germinating wheat. Uch. Zap. Permsk. Gos. Univ. **13**, 33 (1960).

CANNY, M. J., and K. MARKUS: Breakdown of 2,4-dichlorophenoxyacetic acid in shoots and roots. Australian J. Biol. Sci. **13**, 486 (1960).

CARNS, H. R., and F. T. ADDICOTT: The effects of herbicides on endogenous regulator systems. In: Physiology and biochemistry of herbicides. L. J. Audus, ed. Pp. 343-356. New York: Academic Press 1964.

CARROLL, R. B.: Effectiveness of various chemicals in counteracting 2,4-D toxicity to seedlings. Proc. NE Weed Control Conf., pp. 21-22 (1952).

CATLIN, P. B.: Changes in some B vitamins associated with the responses of the Tilton apricot fruit to 2,4,5-trichlorophenoxyacetic acid. Proc. Amer. Soc. Hort. Sci. **74**, 174 (1959).

—, and E. C. MAXIE: Some relations between growth, respiration, and 2,4,5-trichlorophenoxyacetic acid treatment in developing apricot fruits. Proc. Amer. Soc. Hort. Sci. **74**, 159 (1959).

CHACRAVARTI, A. S., D. P. SRIVASTOVA, and K. L. KHANNA: Foliar application of 2,4-D to increase sugar in cane. Sugar J. **18**, 23 (1955).

— — — Soil application of 2,4-D to increase sugar in cane. Current Sci. **25**, 302 (1956).

CHAO, T. F., and H. WANG: Effect of maleic hydrazide and 2,4,5-trichlorophenoxyacetic acid on terminal buds and content of total nitrogen and starch and the rest period of tubers of potatoes. Shih Yen Sheng Wu Hsueh Pao **5**, 515 (1957).

CHKANIKOV, D. I., A. M. MAKEEV, and N. N. PALOVA: Halogen phenoxyacetic acids and oxidative phosphorylation. Khim. v Sel'sk. Khoz **10**, 55 (1964).

CHRISPEELS, M. J., and J. B. HANSON: The increase in ribonucleic acid content of cytoplasmic particles of soybean hypocotyl induced by 2,4-dichlorophenoxyacetic acid. Weeds **10**, 123 (1962).

CLAEYS, R.: Changes in the food reserves of germinating peas as a result of treatment with ammonium 2,4-dichlorophenoxyacetic. Natuurw. Tijdschr. **32**, 115 (1950).

COCORDANO, M., and J. RICARD: Molecular biology of growth. I. Electronic structure and reaction mechanism of chlorinated phenoxyacetic acids. Physiol. Vegetale **1**, 129 (1963).

COLINESE, D. C., J. HALL, and D. A. IBBITSON: Dipole moments of substituted phenoxyacetic acids and phenols. J. Chem. Soc. **1962**, 983.

CONWAY, E.: Effects of γ-(2,4-dichlorophenoxy) butyric acid on sporelings of bracken. Nature **177**, 1088 (1956).

COOKE, A. R.: Influence of 2,4-D on the uptake of minerals from the soil. Weeds **5**, 25 (1957).

CRAFTS, A. S.: The chemistry and mode of action of herbicides. Pp. 52-70. New York: Interscience 1961.

— Herbicide behaviour in the plant. In: Physiology and biochemistry of herbicides. L. J. Audus, ed., pp. 98-102. New York: Academic Press 1964.

CRANE, J. C.: Frost resistance and reduction in drop of injured apricot fruits affected by 2,4,5-trichlorophenoxyacetic acid. Proc. Amer. Soc. Hort. Sci. **64**, 225 (1954).

—, E. D. DeKAZOS, and J. G. BROWN: Effect of 2,4,5-trichlorophenoxyacetic acid on growth, moisture and sugar content of apricot fruits. Proc. Amer. Soc. Hort. Sci. 68, 105 (1956).

CROSBY, D. G.: Metabolites of 2,4-dichlorophenoxyacetic acid (2,4-D) in bean plants. J. Agr. Food Chem. 12, 3 (1964).

CRUZADO, H. J., and T. J. MUZIK: Effect of 2,4-D on the sugar content of sugar cane. Sugar J. 13, 78 (1950).

DIETERMAN, L. J., C. Y. LIN, L. M. ROHRBAUGH, and S. H. WENDER: Accumulation of ayapin and scopolin in sunflower plants treated with 2,4-dichlorophenoxy-acetic acid. Arch. Biochem. Biophys. 106, 275 (1964).

DUNHAM, R. S.: Differential responses in crop plants. In: Plant growth substances. F. Skoog, ed., pp. 195-206. Univ. of Wisc. Press, Madison (1951).

EDGERTON, L. J.: Inactivation of 2,4-dichlorophenoxyacetic acid by apple leaves. Proc. Amer. Soc. Hort. Sci. 77, 22 (1961).

—, and M. B. HOFFMAN: Fluorine substitution affects decarboxylation of 2,4-di-chlorophenoxyacetic acid in apple. Science 134, 341 (1961).

EGOROVA, G. N., A. V. BESHANOV, and M. A. ZARUBINA: Effect of granulated herbicides on the biochemical composition of table beets and Welsh onion tubers. Byul. Vses. Nauchn.-Issled Inst. Zashchity Rast. 1962, 20.

ELSAKOVA, T. N.: Effect of physiologically active compounds on the metabolism of nucleic acids in plants. Biol. Nukleinovago Obmena u Rast., Akad. Nauk SSSR, Bashkirsk Filial, Materialy 2-oi [Vtoroi] Nauchn. Konf., Ufa 1962, 169 (1964).

ELWELL, H. M., and J. E. WEBSTER: Chemical analyses of twigs from oaks sprayed 2,4,5-T and silvex. Proc. 15th NC Weed Control Conf., p. 126 (1958).

ERGLE, D. R., and A. A. DUNLAP: Responses of cotton to 2,4-D. Texas Agr. Expt. Sta. Bull. 713, 5 (1949).

ERICKSON, L. C., C. I. SEELY, and K. H. KLAGES: Effect of 2, 4-D upon the protein content of wheats. J. Amer. Soc. Agron. 40, 659 (1948).

—, R. T. WEDDING, and B. L. BRANNAMAN: Influence of pH on 2,4-D and acetic acid in *Chlorella*. Plant Physiol. 30, 69 (1955).

EPPS, E. A., JR.: Effect of 2, 4-D on growth and yield of cotton. J. Agr. Food Chem. 1, 1009 (1953).

FALUDI, B.: The effect of 2,4-dichlorophenoxyacetic acid (2,4-D) on phospholipids. Ann. Univ. Sci. Budapest Rolando Eotvos Nominatse, Sect. Biol. 5, 63 (1962).

—, and A. FALUDI-DANIEL: Role of the alterations in phosphorus metabolism in resistance to dichlorophenoxyacetic acid. Acta Biol. Acad. Sci. Hung. 11, 43 (1960).

— — The effect of different concentrations of 2,4-dichlorophenoxyacetic acid on the growth, amino acid, and α-keto acid content of potato-tissue cultures. Acta Biol. Acad. Sci. Hung. 8, 273 (1958).

— —, E. KOVACS, and A. BALINT: Action of 2,4-dichlorophenoxyacetic acid (2,4-D) on the phosphorus metabolisms of plants. Biol. Kozlemenyek 7, 7 (1959).

FANG, S. C., and J. S. BUTTS: Carboxy-carbon-14-labeled 3-indoleacetic acid in plants. Plant Physiol. 32, 253 (1957).

— — Plant metabolism. IV. Comparative effects of 2,4-dichlorophenoxyacetic acid and other plant growth regulators on phosphorus metabolism in bean plants. Plant Physiol. 29, 365 (1954a).

— — Plant metabolism. IV. Comparative effects of 2,4-dichlorophenoxyacetic acid and other plant growth regulators on phosphorus metabolism in bean plants. Plant Physiol. 29, 539 (1954b).

FANG, S. C., F. TEENY, and J. S. BUTTS: Influence of 2,4-dichlorophenoxyacetic acid on pathways of glucose utilization in bean-stem tissues. Plant Physiol. 35, 405 (1960).

— — — Effect of 2,4-dichlorophenoxyacetic acid on utilization of labelled acetate by bean leaf and stem tissues. Plant Physiol. 36, 192 (1961).

Fang, S. C., P. Theisen, and J. S. Butts: Metabolic studies of applied indoleacetic acid-1-C[14] in plant tissues as affected by light and 2,4-D treatment. Plant Physiol. **34**, 26 (1959).

Fawcett, C. H., J. M. A. Ingram, and R. L. Wain: The beta oxidation of omega-phenoxy-alkylcarboxylic acids in flax plant in relation to their plant growth regulating activity. Roy Soc. Bot. Proc. **142**, 60 (1954).

Fedorov, N. I., and S. I. Egorova: Effect of growth stimulants on phosphorus and calcium uptake by woody plants. Fiziol. Rast. **10**, 227 (1963).

Fisher, A. M.: Improving the nutritional qualities of tomatoes by means of chemical growth stimulators. Tr. Kazakhsk. Inst. Epidemiol. Mikrobiol. i Gigieny **4**, 565 (1961).

Fites, R. C., F. W. Slife, and J. B. Hanson: Translocation and metabolism of 2,4-D in jimsonweed. Weeds **12**, 180 (1964).

Freed, V. H., and M. L. Montgomery: The metabolism of herbicides by plants and soils. Residue Reviews **3**, 1 (1963).

— —, and M. Kief: The metabolism of certain herbicides by plants—a factor in their biological activity. Proc. 15*th* NE Weed Control Conf., pp. 6-16 (1961 a).

—, F. J. Reithel, and L. F. Remmert: Some physical-chemical aspects of synthetic auxins with respect to their mode of action. In: Plant growth regulation, 4*th* Internat. Conf. on Plant Growth Regulation, pp. 289-303. Ames: Iowa State Univ. Press 1961 b.

Freeland, R. O.: Effects of growth substances on photosynthesis. Plant Physiol. **24**, 621 (1949).

— Effects of 2,4-D and other growth substances on photosynthesis and respiration in *Anacharis*. Bot. Gaz. **111**, 319 (1950).

Freiberg, S. R.: Effects of an exogenous plant growth regulator on proteolytic enzymes of the soybean plant. Science **115**, 674 (1952).

— Effect of growth regulators on ripening, split peel, reducing sugars, and diastatic activity of bananas. Bot. Gaz. **117**, 113 (1955).

—, and H. E. Clark: Effects of 2,4-dichlorophenoxyacetic acid upon the nitrogen metabolism and water relations of soybean plants grown at different nitrogen levels. Bot. Gaz. **113**, 322 (1952).

— — Changes in nitrogen fractions and proteolytic enzymes of soybean plants treated with 2,4-dichlorophenoxyacetic acid. Plant Physiol. **30**, 39 (1955).

French, R. C., and H. Beevers: Respiratory and growth responses induced by growth regulators and allied compounds. Amer. J. Bot. **40**, 660 (1953).

Fults, J. L., and M. G. Payne: The effect of 2,4-D and maleic hydrazide on sprouting, yields and color in Red McClure potatoes. Amer. Potato J. **32**, 451 (1955).

— — Effects of 2,4-dichlorophenoxyacetic acid and maleic hydrazide on free amino acids and proteins in potato, sugar beet, and bean tops. Bot. Gaz. **118**, 130 (1956).

—, R. J. Hay, and M. G. Payne: Nitrate content of Red McClure potatoes unchanged by 2,4-D treatment. Amer. Potato J. **29**, 97 (1952).

Furtick, W. R.: Effects of CIPC, monuron, and 2,4-D herbicides on yield, crude protein, and nitrate content of four forage-crop species. Ph.D. Dissertation. Oregon State College Corvallis (1958).

Gagarina, M. I.: 2,4-Dichlorophenoxyacetic acid in plant tissues. Khim. v Sel'sk. Khoz. **1964** (3), 30.

Galston, A. W., and L. Dalberg: Adaptive formation and physiological significance of indoleacetic acid oxidase. Amer. J. Bot. **41**, 373 (1954).

—, P. Jackson, R. Kaur-Sawhney, N. P. Kefford, and W. J. Meudt: Interactions of auxins with macromolecular constituents of pea seedlings. Colloq. Internat. Centre Natl. Rech. Sci. No. **123**, 251 (1964).

—, and R. Kaur: The intracellular locale of auxin action: an effect of auxin on the physical state of cytoplasmic proteins. In: Plant growth regulation, Fourth

Internat. Conf. on Plant Growth Regulation, pp. 355-362. Ames: Iowa State Univ. Press 1961.

— — An effect of auxins on the heat coagulability of the proteins of growing plant cells. Proc. Nat. Acad. Sci. U.S. **45**, 1587 (1959).

— —, N. MAHESHWARI, and S. C. MAHESHWARI: Pectin-protein interaction as a basis for auxin-induced alteration of protein heat coagulability. Amer. J. Bot. **50**, 487 (1963).

—, and P. J. PENNY: Kinetics of the inhibition of auxin-induced growth in green pea stem sections by ribonuclease and actinomycin D. Plant Physiol. **39**, suppl. *xxx-xxxi* (1964).

GASPARYAN, M. G., and A. A. AVETISYAN: Influence of certain physiologically active substances on enzyme activities in seeds of kyurushma grass. Izvest. Akad. Nauk Armayan. S.S.R., Biol. i Sel'skokhoz. Nauki **11**, 67 (1958).

GLASZIOU, K. T.: Effect of 2,4-D on pectin methylesterase in tobacco pith sections. Nature **181**, 428 (1958).

— The effect of auxins on the binding of pectin methylesterase to cell wall preparations. Australian J. Biol. Sci. **10**, 426 (1957).

—, and S. D. INGLIS: The effect of auxins on the binding of pectin methylesterase to cell walls. Australian J. Biol. Sci. **11**, 127 (1958).

GOLDACRE, P. L.: Action of 2, 4-dichlorophenoxyacetic acid. Australian J. Sci. Research **2B**, 154 (1949).

GRIGSBY, B. H., and C. D. BALL: Some effects of herbicidal spray on the hydrocyanic acid content of leaves of wild black cherry (*Prunus serotina*). Proc. NE 6*th* Weed Control Conf., pp. 327-330 (1952).

GRITSAENKO, Z. M.: Anatomic-physiological changes in plants due to 2,4-D. Nauk Pratsi, Kam'yanets-Podil'sk. Sil's'kogogospodar's Inst. **4**, 76 (1961).

GRUODIENE, J.: Effect of 2,4,5-trichlorophenoxyacetic acid and heteroauxin on the yield and on some physiological processes of potatoes. Lietuvos TSR Aukstuju Mokyklu Mokslo Darbai, Biol. **3**, 31 (1963).

GUNAR, I. I., E. E. KRASTINA, and K. A. BRYUSHKOVA: Effect of 2,4-dichlorophenoxyacetic acid on metabolism in the sunflower at different temperatures. Doklady Akad. Nauk S.S.S.R. **84**, 173 (1952).

GUSTAFSON, F. G.: Riboflavin synthesis by epicotyls of white lupine seedlings grown in sterile culture. Plant Physiol. **30**, 444 (1955).

HAGEN, C. E., C. O. CLAGETT, and E. A. HELGESON: 2,4-Dichlorophenoxyacetic acid inhibition of castor-bean lipase. Science **110**, 116 (1949).

HALL, W. C., and P. W. MORGAN: Auxin-ethylene interrelations. Colloq. Intern. Centre Natil. Rech. Sci. (Paris), No. **123**, 725 (1964).

HANSCH, C., and R. M. MUIR: Electronic effect of substituents on the activity of phenoxyacetic acids. In: Plant growth regulation, Fourth Internat. Conf. on Plant Growth Regulation, pp. 431-443. Ames: Iowa State Press 1961.

— — Ortho effect in plant growth-regulators. Plant Physiol. **25**, 389 (1950).

— —, and R. C. METZENBERG, JR.: Further evidence for a chemical reaction between plant growth-regulators and a plant substrate. Plant Physiol. **26**, 812 (1951).

HANSEN, E.: Effect of 2,4-dichlorophenoxyacetic acid on the ripening of Bartlett pears. Plant Physiol. **21**, 588 (1946).

HANSEN, J. R., and K. P. BUCHHOLTZ: Inactivation of 2,4-D by riboflavin in light. Weeds **1**, 237 (1952).

—, and J. BONNER: Relationship between salt and water uptake in Jerusalem artichoke tuber tissue. Amer. J. Bot. **41**, 702 (1954).

HARRISON, E. M., J. L. FULTS, and M. G. PAYNE: Interaction of 2,4-D, maleic hydrazide and minor-element uptake in potatoes. Amer. Potato J. **35**, 425 (1958).

HATTON, T. T.: Effect of waxes and 2,4,5-trichlorophenoxyacetic acid as post-harvest treatments on Persian limes. Proc. Florida State Hort. Soc. **71**, 312 (1958).

Hay, J. R.: Effect of 2,4-D and 2,3,5-triiodobenzoic acid (TIBA) on an auxin transport mechanism in bean stems. Proc. 12th NC Weed Control Conf., pp. 4-5 (1955).
— Effect of 2,4-dichlorophenoxyacetic acid and 2,3,5-triiodobenzoic acid on the treatment of indoleacetic acid. Plant Physiol. 31, 118 (1956).
—, and K. V. Thimann: Fate of 2,4-D in bean seedlings. I. Recovery of 2,4-D and its breakdown in the plant. Plant Physiol. 31, 382 (1956).
Henderson, J. H. M., and D. C. Desse: Correlation between endogenous auxin and its destruction in vivo by 2,4-dichlorophenoxyacetic acid in plants. Nature 174, 967 (1954).
—, I. H. Miller, and D. C. Desse: Effect of 2,4-D on respiration and on destruction of indoleacetic acid in oat and sunflower tissues. Science 120, 710 (1954).
Hill, K. W.: Protein content of wheat as affected by agronomic practices. Can. J. Plant Sci. 44, 115 (1964).
Hilton, J. L., L. L. Jansen, and H. M. Hull: Mechanisms of herbicidal action. Ann. Rev. Plant Physiol. 14, 353 (1963).
Hoepfner, K. H.: The effect of growth regulators on the growth and inhibitor content of roots. Flora 151, 398 (1961).
Hofmann, E., and B. v. Schmeling: Effect of 2,4-dichlorophenoxyacetic acid on metabolism and enzyme content of plants. Naturwissensch. 40, 23 (1953).
Holley, R. W.: The fate of radioactive 2,4-dichlorophenoxyacetic acid in bean plants. II. A water-soluble transformation product of 2,4-D. Arch. Biochem. Biophys. 35, 171 (1952).
—, F. P. Boyle, and D. B. Hand: The fate of radioactive 2,4-dichlorophenoxyacetic acid in bean plants. Arch. Biochem. 27, 143 (1950).
Hoos, J. W., S. J. Leonard, and B. S. Luh: Effect of 2,4,5-trichlorophenoxyacetic acid spray on organic acids, pectin, and quality of canned apricots. Food Research 21, 571 (1956).
Horowitz, B.: Inhibition of auxiliary growth in Nicotiana rustica plants by chemicals. J. Australian Inst. Agr. Sci. 15, 128 (1949).
Huffaker, R. C., M. D. Miller, and D. S. Mikkelsen: Effects of 2,4-D, iron, and chelate supplements on dark CO_2 fixation in cell-free homogenates of field beans. Crop Sci. 2, 127 (1962).
Humphreys, T. E., and W. M. Dugger, Jr.: Effect of 2,4-dichlorophenoxyacetic acid on pathways of glucose catabolism in higher plants. Plant Physiol. 32, 136 (1957 a).
— — Effect of 2, 4-dichlorophenoxyacetic acid on the respiration of etiolated pea seedlings. Plant Physiol. 32, 530 (1957 b).
— — Effect of 2,4-dichlorophenoxyacetic acid and 2,4-dinctrophenol on the uptake and metabolism of exogenous substrates by corn roots. Plant Physiol. 34, 112 (1959).
— — Metabolism and transport of carbohydrates in plants. Evidence for non-mitochondrial electron transport to oxygen in corn roots treated with 2,4-D. U.S. Atomic Energy Comm. ORO-470 (1961).
Hwang, Y., and C. Shing: Induced parthenocarpy of tomato by 2,4-D. Agr. Research (Formosa) 4, 68 (1954).
Jaworski, E. G., and J. S. Butts: Studies in plant metabolism. II. The metabolism of C^{14} labeled 2,4-dichlorophenoxyacetic acid on bean plants. Arch. Biochem. Biophys. 38, 207 (1952).
—, S. C. Fang, and V. H. Freed: Plant metabolism. V. Metabolism of radioactive 2,4-D in etiolated bean plants. Plant Physiol. 30, 272 (1955).
Johnson, E. J., and A. R. Colmer: Relation of structure of 2,4-dichlorophenoxyacetate to its mode of action as an auxin. Nature 180, 1365 (1957).
— — Relationship of magnesium ion and molecular structure of 2,4-dichlorophenoxyacetic acid and some related compounds to the inhibition of the respiration of Azotobacter vinelandii. Plant Physiol. 33, 99 (1958).

OHNSTON, A. M.: Relation of structure to activity in plant growth-regulating compounds. II. Growth-regulating activity of substituted amides of 2,4-dichlorophenoxyacetic acid. Biochem. J. **82**, 425 (1962).

ULG, A., and M. COCORDANO: Electronic structure and auxinic activity of chlorinated derivatives of phenoxyacetic acid. Compt. rend. **254**, 1070 (1962).

KADLUBOWSKI, R., and H. SKALECKA: Effect of β-indoleacetic acid and sodium 2,4-dichlorophenoxyacetate on the saponin content in soapwort (*Saponaria officinalis*) and foxglove (*Digitalis purpurea*). Acta Physiol. Polon. **12**, 583 (1961).

KAMAL, A: Effect of some insecticides and 2,4-D ester on the nitrogen fractions of Bartlett pear leaf and stem tissues. Ph.D. Dissertation. Washington State Univ., Pullman (1960).

KANDLER, O.: The effect of 2,4-D, α napthaleneacetic acid, sodium fluoride, and vitamin B upon the metabolism of severed maize roots grown in culture. Planta **42**, 304 (1953).

KASPERLIK, H.: Comparative physiological investigations, of 2,4-D-sensitive and -resistant plants. Flora **142**, 307 (1955).

KECK, K., and O. HOFFMAN-OSTENHOF: The chemically induced chromatin escape from plant cell nuclei. Expt. Cell Research **7**, 111 (1954).

KELLY, S., and G. S. AVERY, JR.: Effect of 2,4-dichlorophenoxyacetic acid and other physiologically active substances on respiration. Amer. J. Bot. **36**, 421 (1949).

— — The age of pea tissue and other factors influencing the respiratory response to 2,4-dichlorophenoxyacetic acid and dinitro compounds. Amer. J. Bot. **38**, 1 (1951).

KENT, N. L., and J. B. HUTCHINSON: The effect of MCPA and 2,4-D herbicides on the quality of spring oats for milling and provender feeding. Ann. Applied Biol. **45**, 481 (1957).

KEY, J. L.: Biochemical effects of 2,4-dichlorophenoxyacetic acid on plants. Ph.D. Dissertation. Univ. of Illinois, Urbana (1959).

— Changes in ascorbic acid metabolism associated with auxin-induced growth. Plant Physiol. **37**, 349 (1962).

— 2,4-D-induced changes in ribonucleic acid metabolism in excised corn mesocotyl tissue. Weeds **11**, 177 (1963).

— Ribonucleic acid and protein synthesis as essential processes for cell elongation. Plant Physiol. **39**, 365 (1964 a).

— RNA synthesis and the control of expansive growth of excised soybean hypocotyl. Plant Physiol. **39**, suppl. p. *xxx* (1964 b).

—, and D. S. GALITZ: Growth inhibitor in immature soybean seeds and 2,4-D-sprayed soybean seedlings. Science **130**, 1340 (1959).

—, and J. B. HANSON: Effect of 2,4-D treatment on the respiratory metabolism of etiolated soybean seedlings. Proc. 15*th* NC Weed Control Conf., pp. 3-4 (1958).

— — Effects of 2,4-dichlorophenoxyacetic acid on soluble nucleotides and nucleic acid of soybean seedlings. Plant Physiol. **36**, 145 (1961).

— —, and R. F. BILS: Effect of 2,4-dichlorophenoxyacetic acid application on activity and composition of mitochondria from soybeans. Plant Physiol. **35**, 177 (1960).

—, and J. C. SHANNON: Enhancement by auxin of ribonucleic acid synthesis in excised soybean hypocotyl tissue. Plant Physiol. **39**, 360 (1964).

—, and F. WOLD: Some effects of 2,4-dichlorophenoxyacetic acid on the oxidation-reduction state of soybean seedlings. J. Biol. Chem. **236**, 549 (1961).

KHUBUTIYA, R. A.: The transformation of 2,4-dichlorophenoxyacetic acid in crops. Soobschcheniya Akad. Nauk Gruzin. S.S.R. **21**, 313 (1958).

— Effect of 2,4-D on processes of oxidative phosphorylation. Agrobiologiya **1**, 139 (1959).

KLAEMBT, H. D.: Induction of growth and metabolism of growth substances in wheat coleoptile. III. Products of the metabolism of napthaleneacetic acid and

2,4-dichlorophenoxyacetic acid and comparison with those of indoleacetic acid and benzoic acid. Planta **57**, 339 (1961).

KLINGMAN, D. L.: Effects of varying rates of 2,4-D and 2,4,5-T at different stages of growth on winter wheat. Agron. J. **45**, 606 (1953).

KONDO, H., A. KAGAWA, and K. HAYASHI: Electrophoretic proteins of vegetable tissues. Nippon Nôgei-Kagaku Kaishi **31**, 433 (1957).

KOULA, V., and M. KRAMLOVA: Determination of the secondary effects of esters and salts of substituted phenoxyacetic acids, applied as cold aerosols, on the content of biogenic and trace elements of weeds and agricultural plants. Ustav Vedecko-tech. Inform. Min. Zemedel., Lesniko Vodniko Hospodarstvi Rostlinna Vyroba **10**, 451 (1964).

KOZINKA, V.: Changes of transpiration intensity caused by 2-methyl-4-chlorophen-oxyacetic acid. Biologia **19**, 809 (1964).

KUNERT, G.: Influence of some herbicides on the lipase activity of *Aspergillus niger.* Naturwissensch. **46**, 603 (1959).

KVAMME, O. J., C. O. CLAGETT, and W. B. TREUMANN: Kinetics of the action of the sodium salt of 2,4-dichlorophenoxyacetic acid on germ lipase of wheat. Arch. Biochem. **24**, 321 (1949).

LADONIN, V. F.: Effect of 2,4-D esters on certain physiological processes in plants. Tr. Vses. Nauchn.-Issled. Inst. Udobr. i Agropochvoved. **36**, 159 (1960).

— Influence of several herbicides on the nitrogen content of plants. Primenenie Gerbitsidov v Sel'sk. Khoz., Upr. Nauk, Propagandy i Vnedreniya Peredovogo Optyta, Min. Sel'sk. Khoz. SSSR **1962**, 221.

LAWRENCE, J. M., and T. J. MUZIK: Comparison of changes in nitrogenous constituents following uprooting or 2,4-D treatments of bean plants. Northwest Sci. **36**, 39 (1962).

LEAFE, E. L.: Metabolism and selectivity of plant-growth regulator herbicides. Nature **193**, 485 (1962).

LEARY, J. B., and C. O. CLAGETT: 2,4-Dichlorophenoxyacetic acid inhibition of plant enzymes. Potato phosphorylase. Proc. N. Dakota Acad. Sci. **5**, 47 (1951).

LEOPOLD, A. C., and F. S. GUERNSEY: A theory of auxin action involving coenzyme A. Proc. Nat. Acad. Sci. U.S. **39**, 1105 (1953).

LIBBERT, E.: The indoleacetic acid-forming enzyme system of pea plants. Z. Bot. **48**, 365 (1960).

LIKHOLAT, T. V.: Effect of 2,4-D on accumulation of energy-rich phosphates in plants of various systematic position. Fiziol. Rast. **11**, 1070 (1964).

LIN, C., and J. L. KEY: Auxin-induced growth and its relationship to ascorbic acid metabolism. Plant Physiol. **39**, suppl. p. *iii* (1964).

LINDEN, G.: The mode of action of 2,4-dichlorophenoxyacetic acetic acid (2,4-D) in connection with the respiratory intensity of the plant. Beitr. Biol. Pflanz. **29**, 197 (1952).

LINDER, P. J., J. W. BROWN, and J. W. MITCHELL: Movement of externally applied phenoxy compounds in bean plants in relation to conditions favoring carbohydrate translocation. Bot. Gaz. **110**, 628 (1949).

LINSCOTT, D. L.: Degradation of 4-(2-,4-dichlorophenoxy)-butyric acid [4-(2,4-DB)] in plants. J. Agr. Food Chem. **12**, 7 (1964).

LIVINGSTON, C., M. G. PAYNE, and J. L. FULTS: Effects of maleic hydrazide and 2,4-dichlorophenoxyacetic acid on the free amino acids of sugar beets. Bot. Gaz. **116**, 148 (1954).

LOCKHART, J. A., and R. L. WEINTRAUB: Influence of 2,4-dichlorophenoxyacetic acid on auxin content of bean seedlings. Amer. J. Bot. **44**, 424 (1957).

LOTLIKER, P. D.: Effects of herbicides on oxidative phosphorylation in mitochondria from cabbage *Brassica oleracea.* Ph.D. Dissertation. Oregon State College, Corvallis (1960).

LOUSTALOT, A. J., M. J. MORRIS, J. GARCIA, and C. PAGAN: 2,4-D affects phosphorus metabolism. Science **118**, 627 (1953).

LUCKWILL, L. C., and C. P. LLOYD-JONES: Metabolism of plant growth regulators. I. 2,4-Dichlorophenoxyacetic acid (2,4-D) in leaves of red and of black current. Ann. Applied Biol. 48, 613 (1960 a).

— — Metabolism of plant growth regulators. II. Decarboxylation of 2,4-dichlorophenoxyacetic acid in leaves of apple and strawberry. Ann. Applied Biol. 48, 626 (1960 b).

LUECKE, R. W., C. L. HAMNER, and H. M. SELL: Effect of 2,4-dichlorophenoxyacetic acid on the content of thiamine, riboflavin, nicotinic acid, pantothenic acid, and carotene in stems and leaves of red kidney bean plants. Plant Physiol. 24, 546 (1949).

LYNN, G. E., and K. C. BARRONS: The hydrocyanic acid content of wild cherry leaves sprayed with a brush killer containing low-volatile esters of 2,4-D and 2,4,5-T. Proc. 6th NE Weed Control Conf. pp. 331-333 (1952).

MACEY, M.: The effect of some plant growth substances on the esterase activity of whole tissue slices. Proc. 2nd Australian Weed Conf. (1960).

MACIEJEWSKA-POTAPCZYKOWA, W.: Effect of 2,4-D on the chlorophyll content of higher plants. Bull. soc. sci. et lettres Lodz Class III, 4, 1 (1953).

— Effect of 2,4-D on absorption and evaporation of water by plants. Zeszyty Nauk. Univ. Lodz, 3, 95 (1957).

MACLEOD, D. G.: The effect of MCPA on the amino acid and sugar composition of Vicia faba leaves. Weed Res. 4, 275 (1964).

MALISAUSKIENE, V.: Action of 2,4-D on the physiological processes in oats and on the yield. Lietuvos TSR Mokslu Akad. Darbai, Ser. B. 1, 107 (1959).

— Effect of herbicide 2,4-D on photosynthesis, crop yield, and quality and weight of grains. Primenenie Gerbitsidov i Stimulyatorov Rastenii Sb. 1961, 27.

— Effect of 2,4-D on photosynthesis and chlorophyll content of mono- and dicotyledons. Fiziol. ir Biochem. Klausimai Lectuvos TSR Mokslu Akad. Botan. Inst. 1962, 217.

MARTH, P. C.: Effect of growth regulators on the retention of color in green sprouting broccoli. Proc. Amer. Soc. Hort. Sci. 60, 367 (1952).

—, and J. W. MITCHELL: Apparent antagonistic effects of growth regulators. Bot. Gaz. 110, 514 (1949).

MASHTAKOV, S. M., A. Z. DENISOVA, Z. I. PARADOVSKAYA, and Z. P. PARSHAKOVA: Effect of the sodium salt of 2-methyl-4-chlorophenoxyacetic acid on content of nucleic acid in corn. Dokl. Akad. Nauk Belorussk, SSR 8, 677 (1964).

—, and I. I. PAROMCHICK: Twenty-four hour cycle of photosynthesis in plants resistant to regulator herbicides. Dokl. Akad. Nauk Belorussk. SSR 6, 801 (1962).

— —, and K. S. TALANOVA: Photosynthesis and respiration of hybrids and varieties of corn under the effect of the sodium salts of 2,4-D and 2M-4C. Vesti Akad. Navuk Belarusk. SSR, Ser. Biyal. Navuk No. 2, 43 (1962).

MASUDA, Y.: Effect of ribonuclease on the ribonucleic acid of Avena coleoptile cells. J. Inst. Polytech. Osaka City Univ. 10D, 1 (1959).

MAXIE, E. C., M. U. BRADLEY, and B. J. ROBINSON: The accumulation of C14 from carboxyl-labeled 2,4,5-trichlorophenoxyacetic acid in fruit of Tilton apricot. Proc. Amer. Soc. Hort. Sci. 81, 137 (1962).

—, and J. C. CRANE: Some metabolic effects of 2,4,5-trichlorophenoxyacetic acid on Tilton apricot fruits. Proc. Amer. Soc. Hort. Sci. 68, 113 (1956).

McILRATH, W. J., and D. R. ERGLE: Further evidence of persistance of the 2,4-D stimulus in cotton. Plant Physiol. 28, 693 (1953).

— —, and A. A. DUNLAP: Persistance of 2,4-D stimulus in cotton plants with reference to its transmission to the seed. Bot. Gaz. 112, 511 (1951).

McNEW, G. L., and O. L. HOFFMANN: Growth-regulant, herbicidal, and physical properties of 2,4-D and related compounds. Iowa State College J. Sci. 24, 189 (1950).

MÉNORET, Y., and G. MOREL: Influence of 2,4-dichlorophenoxyacetic acid on the

106   DONALD PENNER and FLOYD M. ASHTON

metabolism of the free amino acids of carrot tissue cultures. Compt. rend. **246**, 2652 (1958).

MEREZHINSKII, Y. G.: The effect of derivatives of phenoxyacetic acid on corn and weeds. Nauchn. Tr., Ukr. Nauckn.-Issled. Inst. Fiziol. Rast., Ukr. Akad. Sel's-kokhoz. Nauk **23**, 132 (1962).

MILLER, I. H., JR., and R. H. BURRIS: Effect of plant-growth substances upon oxidation of ascorbic and glycolic acids by cell-free enzymes from barley. Amer. J. Bot. **38**, 547 (1951).

MITCHELL, J. W., and J. W. BROWN: Effect of 2,4-dichlorophenoxyacetic acid on the readily available carbohydrate constituents in annual morning-glory. Bot. Gaz. **107**, 120 (1945).

MITCHELL, J. E., R. H. BURRIS, and A. J. RIKER: Inhibition of respiration in plant tissue by callus-stimulating substances and related chemicals. Amer. J. Bot. **36**, 368 (1949).

MITCHELL, J. W., and P. C. MARTH: Effects of 2,4-dichlorophenoxyacetic acid on the ripening of detached fruits. Bot. Gaz. **106**, 199 (1944).

MOEWUS, F.: Action of 2,4-D on diaminating enzymes. Congress Internat. Bot. Paris, Rapps. et communs. 8, section 11/12, 149 (1964).

MOLOTKOVSKII, G. K., and N. I. VOLOTOVSKAYA: Activation of lipase by some growth stimulants. Doklady Akad. Nauk S.S.S.R. **70**, 117 (1950).

MONSTVILAITE, Y.: Effect of systematic use of the herbicide 2,4-dichlorophenoxyacetic acid (2,4-D) on the yield of spring grain crops. Sb. Dokl. Nauckn. Konf. po Zaskchite Rast., Tartu **1960**, 374 (1962).

MOREL, G., and S. DÉMÉTRIADES: Action of growth regulators on the oxidizing activity of Jerusalem artichoke tissues. Année biol. **59**, 227 (1955).

MORELAND, D. E., and K. L. HILL: Interference of herbicides with the Hill reaction of isolated chloroplasts. Weeds **10**, 229 (1962).

MORGAN, P. W., and W. C. HALL: Metabolism of 2,4-D by cotton and grain sorghum. Weeds **11**, 130 (1963).

MORRÉ, D. J., and B. J. ROGERS: The fate of long chain esters of 2,4-D in plants. Weeds 8, 436 (1960).

MORTON, H. L., and R. E. MEYER: Absorption, translocation and metabolism of 2,4,5-T by mesquite seedlings. Plant Physiol. **37**, suppl. *xxiv-xxv* (1962).

MUIR, R. M., and C. HANSCH: Mechanism of action of growth regulators. Plant Physiol. **28**, 218 (1953).

MUZIK, T. J., and J. M. LAWRENCE: Amino-acid metabolism in bean roots as affected by drying and 2,4-D. Proc. 16*th* W Weed Control Conf., p. 31 (1958).

— — Separation of specific and nonspecific effects of chemicals on plants. Nature **183**, 482 (1959).

—, and J. W. WHITWORTH: Growth-regulating chemicals persist in plants: qualitative bioassay. Science **140**, 1212 (1963).

NAKAZAWA, K.: Physiological action of phytohormone. Igaku to Seibutsugsku **16**, 188 (1950).

NANCE, J. F.: Inhibition of salt accumulation in excised wheat roots by 2,4-dichlorophenoxyacetic acid. Science **109**, 174 (1949).

—, and L. W. Cunningham: Acetaldehyde accumulation in excised wheat roots by plant growth substances. Science **112**, 170 (1950).

NEELY, W. B., C. D. BALL, C. L. HAMNER, and H. M. SELL: Effect of 2,4-dichlorophenoxyacetic acid on the α- and β-amylase activity in the stems and leaves of red kidney bean plants. Science **111**, 118 (1950 a).

— — — — Effect of 2,4-dichlorophenoxyacetic acid on the invertase, phosphorylase, and pectinmethoxylase activity in the stems and leaves of the red kidney bean plants. Plant Physiol. **25**, 525 (1950 b).

NYLUND, R. E.: The use of 2,4-D to intensify the skin color of Pontiac potatoes. Amer. Potato J. **33**, 145 (1956).

OKUNTSEV, M. M., and G. D. ZYRYANOVA: Effect of herbicides on plant metabolism. Izvest. Tomsk-Otdela Vsesoyuz. Botan. Obshchestva 4, 87 (1959).

OMROD, D. C., and W. A. WILLIAMS: Phosphorus metabolism of *Trifolium hirtum*
as affected by 2,4-dichlorophenoxyacetic acid (2,4-D) and gibberellic acid. Plant
Physiol. 35, 81 (1960).

OOTA, Y.: RNA in developing plant cells. Ann. Rev. Plant Physiol. 15, 17 (1964).

OSBORNE, D. J., and M. HALLAWAY: Auxin control of protein levels in detached
autumn leaves. Nature 188, 240 (1960).

— — The role of auxins in the control of leaf senescence. Some effects of local
applications of 2,4-dichlorophenoxyacetic acid on carbon and nitrogen metabolism.
In: Plant growth regulation. 4th Internat. Conf. on Plant Growth Regulation,
pp. 329-338. Ames: Iowa State Univ. Press 1961.

— — The auxin, 2,4-dichlorophenoxyacetic acid, as a regulator of protein synthesis
and senescence in detached leaves of *Prunus*. New Phytologist 63, 334 (1964).

OTANI, K.: Effect of 2,4,5-trichlorophenoxypropionic acid on the maturation of
persimmons. Tokyo Nogyo Daigaku Shuho 5, 53 (1959).

PALEG, L. G., and R. M. MUIR: Surface activity as related to physiological activity
of plant growth regulators. Plant Physiol. 27, 285 (1952).

PARIS, P.: 2,4-D action on root growth and phosphorus uptake in herbicide-sensitive
plants. Pubbl. univ. Cattolica S. Cuore. Ann. fac. Agrar. 78, 29 (1960).

PAYNE, M. G., and J. L. FULTS: The effect of maleic hydrazide and 2,4-D on reducing and sucrose of Red McClure potatoes. Amer. Potato J. 32, 144 (1955).

— —, and R. J. HAY: Free amino acids in potato tubers altered by 2,4-D treatment of plants. Science 114, 204 (1951).

— — — The effect of 2,4-D treatment on free amino acids in potato tubers. Amer.
Potato J. 29, 142 (1952).

— — —, and C. H. LIVINGSTON: Protein content and specific gravity of Red
McClure potatoes increased by 2,4-D treatment. Amer. Potato J. 30, 46 (1953).

PERUANSKII, Y. V.: Influence of derivatives of 2,4-D and 2,4,5-T on the accumulation of nucleic acids and other compounds of phosphorus in soybean leaves.
Vestn. Sel'skokhoz. Nauki, Min. Sel'sk. Khoz. Kaz. SSR 7, 72 (1964).

PICKETT, W. F., A. S. FISH, JR., and K. W. SHAN: Influence of certain organic spray
materials on the photosynthetic activity of peach and apple foliage. Proc. Amer.
Soc. Hort. Sci. 57, 111 (1951).

PILET, P. E., and W. WURGLER: Variations in starch content following treatment
with auxins. Bull. soc. vaudoise sci-nat. 65, 397 (1953).

PLAYER, M. A.: Effects of some growth-regulating substances on the transpiration
of *Zea mays* and *Ricinus communis*. L. Plant Physiol. 25, 469 (1950).

POPOV, K. I., K. DILOV., and P. PETKOV: The effect of 2,4-D on the transpiration
in apricot and apple trees. Izvest. Inst. Rastenievudstvo. Bulgar. Akad. Nauk 2,
181 (1954).

RAKITIN, Y. V., and A. D. POTAPOVA: Effect of 2,4-D and isopropyl 3-chlorophenyl-
carbamate on transpiration and some colloidal properties of the protoplasm.
Doklady Akad. Nauk S.S.S.R. 126, 688 (1959 a).

— — Effect of herbicides on respiration and photosynthesis in oats and sunflowers.
Doklady Akad. Nauk S.S.S.R. 126, 1371 (1959 b).

— — Penetration of herbicides in plants and their effects on phosphorus uptake.
Fiziol. Rastenii, Akad. Nauk S.S.S.R. 6, 614 (1959 c).

—, K. L. POVOLOTSKAYA, and D. M. SEDENKO: Some changes in metabolism of
flowers and plants of tomato after action of 2,4-D and 2,4,5-T. Fiziol. Rastenii,
Akad. Nauk S.S.S.R. 3, 297 (1956).

—, and V. A. ZEMSKAYA: Influence of 2,4-D on nitrogen metabolism in oat and
bean plants. Fiziol. Rastenii, Akad. Nauk S.S.S.R. 5, 97 1958).

RAVAZZONI, C.: Action of 2,4-dichlorophenoxyacetic acid on the phosphate enzyme
systems of plants. Ricerca sci. 19, 376 (1949).

108   DONALD PENNER and FLOYD M. ASHTON

RAVAZZONI, C.: Action of growth substance and of herbicides on the enzyme systems of plants. Farm. sci. e tec. **6**, 588 (1951).
—, and R. VALERIO: Lipase. III. Influence of some cholesterol esters with plant growth substance on lipase activity of *Ricinus*. Giorn. biochim. **5**, 37 (1956).
REBSTOCK, T. L., C. L. HAMNER, C. D. BALL, and H. M. SELL: Effect of 2,4-dichlorophenoxyacetic acid on proteolytic activity of red kidney bean plants. Plant Physiol. **27**, 639 (1952).
— —, and H. M. SELL: Influence of 2,4-dichlorophenoxyacetic acid on the phosphorus metabolism of cranberry bean plants (*Phaseolus vulgaris*). Plant Physiol. **29**, 490 (1954).
REICH, E., R. M. FRANKLIN, A. J. SHATKIN, and E. C. TATUM: Effect of actinomycin D on cellular nucleic acid synthesis and virus production. Science **134**, 556 (1961).
REPP, G.: The selective action of 2,4-D compounds. II. The action of 2,4-D on water content and percentage dry matter. Pflanzensch. Ber. **12**, 181 (1954).
RHODES, A.: The influence of the plant growth-regulator, 2 methyl-4-chlorophenoxyacetic acid, on the metabolism of carbohydrate, nitrogen, and minerals in *Solanum lycopersicum* (tomato). J. Expt. Bot. **3**, 129 (1952).
—, W. G. TEMPLEMAN, and M. N. THURSTON: Effect of the plant-growth regulator 4-chloro-2-methylphenoxyacetic acid on the mineral and nitrogen contents of plants. Ann. Bot. **14**, 181 (1950).
RIRIE, D., D. S. MIKKELSEN, and R. S. BASKETT: Effects of maleic hydrazide and 2,4-D on sugar-beet growth and sugar content in certain field experiments. Proc. Amer. Soc. Sugar Beet Technol. **7**, 86 (1952).
ROSS, M., and F. B. SALISBURY: The effect of herbicides on high-energy phosphate levels. Proc. 19*th* W Weed Control Conf. pp. 90-94 (1962).
SABUROVA, P. V., and L. N. LOGINOVA: Use of radioactive isotopes for studying the mechanism of the selective action of the herbicide 2,4-D. Materially Simpoziuma po Primenemiyu Biofiz. v Obl. Zaskchity Rast., Leningrad, Sb. **1961**, 50.
SAID, H., and M. I. NAGUIB: The effect of 2,4-dichlorophenoxyacetic acid on respiration and carbohydrate metabolism of starved and sucrose-fed carrot-root slices. Proc. Egypt. Acad. Sci. **11**, 27 (1955).
—, and E. H. YOUSSEF: The effect of 2,4-dichlorophenoxyacetic acid on respiration and nitrogen metabolism of radish-root slices. Proc. Egypt. Acad. Sci. **11**, 49 (1955).
SATAROVA, N. A.: Effect of nucleic acids, adenine, and 2,4-D on the growth of gladiolus tissue cultures. Fiziol. Rastenii, Akad. Nauk S.S.S.R. **8**, 301 (1961).
SCIUCHETTI, L. A.: Effects of atropine and 2,4-dichlorophenoxyacetic acid on the growth and alkaloid formation in members of the Solanaceae. Ph.D. Dissertation. Univ. of Washington, Seattle (1957).
SELL, H. M., C. L. HAMNER, T. L. REBSTOCK, L. E. WELLER, A. F. MILLER, and H. N. FUKUI: The effect of di- and trichlorophenoxyacetic acids on the composition of bean plants (*Phaseolus vulgaris*). Michigan State Univ., Agr. Expt. Sta. Quart. Bull. **40**, 44 (1957).
—, R. W. LUECKE, B. M. TAYLOR, C. L. HAMNER: Changes in chemical composition of the stems of red kidney bean plants treated with 2,4-dichlorophenoxyacetic acid. Plant Physiol. **24**, 295 (1949).
SHANNON, J. C.: The effect of 2,4-dichlorophenoxyacetic acid on the growth and ribonuclease content of seedling tissues. Ph.D. Dissertation. Univ. of Illinois, Urbana (1963).
—, J. B. HANSON, and C. M. WILSON: Ribonuclease levels in the mesocotyl tissue of *Zea mays* as a function of 2,4-dichlorophenoxyacetic acid application. Plant Physiol. **39**, 804 (1964).
SHAW, W. C., J. L. HILTON, D. E. MORELAND, and L. L. JENSEN: Herbicides in plants. In: Nature and fate of chemicals applied to soils, plants, and animals. U.S. Dept. of Agr. ARS 20-29, pp. 119-133 (1960).

—, C. J. WILLARD, and R. L. BERNARD: Effect of 2,4-dichlorophenoxyacetic acid on wheat, oats, and barley. Ohio Agr. Expt. Sta., Research Bull. No. 761, 1 (1955).

SHELLENBERGER, J. A., W. M. PHILLIPS, J. A. JOHNSON, and B. S. MILLER: Quality of hard red winter wheat as affected by 2,4-D spray applications. Cereal Chem. 27, 162 (1950).

SILBERGER, J., and F. SKOOG: Changes induced by indoleacetic acid in nucleic acid content and growth of tobacco pith tissue. Science 118, 443 (1953).

SIVORI, E. M.: Inhibition of growth stimulation by 2,4-D and the necessity of phosphorus and potassium for its action. Idia 62, 13 (1953).

—, and F. K. CLAVER: Influence of 2,4-D on oxidation and reduction enzymes. Rev. Argentina Agron. 17, 1 (1950).

SKOOG, F.: Plant growth substances. Madison, Wisc.: Univ. Press 1951.

— Substances involved in normal growth and differentiation of plants. Brookhaven symposia in biology 6 (BNL) 258, 1 (1954).

SKRIPITSYNA, N. E.: Effect of 2,4-dichlorophenoxyacetic acid on tomatoes with variation of mineral diet. Doklady Akad. Nauk. S.S.S.R. 75, 457 (1950).

SLIFE, F. W., J. L. KEY, S. YAMAGUCHI, and A. S. CRAFTS: Penetration, translocation, and metabolism of 2,4-D and 2,4,5-T in wild and cultivated cucumber plants. Weeds 10, 29 (1962).

SMITH, F. G.: Effect of 2,4-dichlorophenoxyacetic acid on the respiratory metabolism of bean stem tissue. Plant Physiol. 23, 70 (1948).

—, C. L. HAMNER, and R. F. CARLSON: Changes in food reserves and respiratory capacity of bindweed tissues accompanying herbicidal action of 2,4-dichlorophenoxyacetic acid. Plant Physiol. 22, 58 (1947).

SMITH, L. H.: Effect of 2,4-dichlorophenoxyacetic acid on seedling development and uptake and distribution of calcium and phosphorus in barley. Ph.D. Dissertation. Mich. State Univ., East Lansing (1959).

—, and C. M. HARRISON: Effect of 2,4-dichlorophenoxyacetic acid on seedling development and uptake and distribution of Ca and P in barley. Crop Sci. 2, 31 (1962).

SORASUCHART, P., J. A. SMITH, and G. R. PATERSON: The effect of 2,4-D and gibberellic acid on the glycoside content of Ornithogalum umbellatum. Can. Pharm. J., Sci. Sect. 95, 496 (1962).

SOUTHWICK, F. W.: Effect of some growth-regulating substance on the rate of softening, respiration, and soluble solids content of peaches and apples. Proc. Amer. Soc. Hort. Sci. 47, 84 (1946).

STAHLER, L. M., and E. I. WHITEHEAD: Effect of 2,4-D on potassium nitrate levels in leaves of sugar beets. Science 112, 749 (1950).

STALLWORTH, H.: Some effects of 2,4-dichlorophenoxyacetic acid on sweet corn (Zea mays rugosa L.) with emphasis on yield, tillering, root development, and exudation of electrolytes from roots and stems. Ph.D. Dissertation. Univ. of Illinois, Urbana (1962).

STAN, S.: Morphological, physical, and biochemical research on tomato plants treated with growth stimulating substances. Analele Inst. Cercetari Agron, Acad. Rep. Populare Romine Ser. C 27, 9 (1960).

STENLID, G., and K. SADDIK: Effect of some growth regulators and uncoupling agents upon oxidative phosphorylation in mitochondria of cucumber hypocotyls. Physiol. Plantarum 15, 369 (1962).

STEVENS, V. L., J. S. BUTTS, and S. C. FANG: Effects of plant-growth regulators and herbicides on metabolism of C^{14}-labeled acetate in pea root tissues. Plant Physiol. 37, 215 (1962).

SUDI, J.: Induction of the formation of complexes between aspartic acid and indolyl-3-acetic acid or 1-naphthaleneacetic acid by other carboxylic acids. Nature 201, 1009 (1964).

SUDI, J., G. JOSEPOVITS, and G. MATOLCSY: Studies with chlorophenoxy derivatives on the possibility of selective herbicidal action based on the precursor principle. Novenyved. Tud. Tanacskozas Kolzlemen., Budapent 2, 401 (1961).

SWANSON, C. R.: Metabolic fate of herbicides in plants. Crops Research, U.S. Dept. of Agr. ARS 34-66, pp. 1-36 (1965).

—, S. B. HENDRICKS, V. K. TOOLE, and C. E. HAGEN: Effect of 2,4-dichlorophenoxy-acetic acid and other growth-regulators on the formation of red pigment in Jerusalem artichoke tuber tissue. Plant Physiol. 31, 315 (1956).

—, and W. C. SHAW: The effect of 2,4-D on the hydrocyanic acid and nitrate content of Sudan grass. Agron. J. 46, 418 (1954).

SWENSON, G., and H. BURSTROM: On the influence of auxins on salt and water uptake. Physiol. Plantarum 13, 846 (1960).

SWITZER, C. M.: Effects of herbicides and related chemicals on oxidation and phosphorylation by isolated soybean mitochondria. Plant Physiol. 32, 42 (1957).

SYNERHOLM, M. E., and P. W. ZIMMERMAN: Preparation of a series of omega-(2,4-dichlorophenoxy)-aliphatic acids and some related compounds with a consideration of their biochemical role as plant growth regulators. Contrib. Boyce Thompson Inst. 14, 369 (1947).

SZABO, S. S.: The hydrolysis of 2,4-D esters by bean and corn plants. Weeds 11, 292 (1963).

TAKAHASHI, T., and M. NAKAYAMA: Influences of hormone spray on the growth and pigment content of tomato fruit. Shinshu Daigaku Nogakubu. Kiyo 2, 151 (1960).

TAYLOR, D. L.: Effects of 2,4-dichlorophenoxyacetic acid on gas exchange of wheat and mustard seedlings. Bot. Gaz. 109, 162 (1947).

THIMANN, K. V.: The role of ortho-substitution in the synthetic auxins. Plant Physiol. 27, 392 (1952).

THOMAS, E. W., B. C. LOUGHMAN, and R. G. POWELL: Metabolic fate of 2,4-dichlorophenoxyacetic acid in the stem of *Phaseolus vulgaris*. Nature 204, 884 (1964a).

— — — Metabolic fate of some chlorinated phenoxyacetic acids in the stem tissue of *Avena sativa*. Nature 204, 286 (1964b).

TITOVA, O. V.: The effect of treatment of the soil and seeds with 2,4-D upon the nitrogen metabolism of wheat seedlings. Uch. Zap. Permsk. Gos. Univ. 18, 33 (1961).

—, and T. P. MIKHAILOVA: The effect of pretreatment of the soil with 2,4-D on certain characteristics of carbohydrate metabolism in wheat. Uch. Zap. Permsk. Gos. Univ. 18, 41 (1961).

TOMISEK, A. J., M. R. REID, W. A. SHORT, and H. E. SKIPPER: The photosynthetic reaction. III. The effects of various inhibitors upon growth and carbonate fixation in *Chlorella pyrenoidosa*. Plant Physiol. 32, 7 (1957).

TOMIZAWA, C.: Effects of 2,4-D and CMU on phosphorus metabolism in soybean plants. Nôgyô Gijutsu Kenkyûjo Hôkoku Ser. C., No. 6, 103 (1956).

—, and H. KOIKE: Herbicides. Bull. Nat. Inst. Agr. Sci., Ser. C, No. 4, 25 (1954).

TORCHINSKAYA, V. M.: Effects of 2,4-dichlorophenoxyacetic acid on growth, multiplication, and nitrogen excange in some weeds. Biol. Zbirnik, L'viv. Derzhav. Univ. im. Ivana Franka, No. 8, 141 (1958a).

— Effect of 2,4-dichlorophenoxyacetic acid on metabolism of nitrogenous substances in lupine sprouts and wilting makhorka leaves. Doklady Akad. Nauk S.S.S.R. 120, 1144 (1958b).

TSAO, D. P. N., and H. W. YOUNGKEN, JR.: Certain chemical plant growth regulators and alkaloid formation in *Datura stramonium*. J. Amer. Pharm. Assoc. 38, 112 (1949).

TSITOVICH, I. K., F. P. GORODKOV, and N. G. MISHUSTINA: Effect of 2,4-D on the enzyme activity of dicotyledonous and cereal plants. Trudy Kuban. Sel'-skokhoz. Inst., No. 2, 127 (1955).

TUKEY, H. B., and C. L. HAMNER: Form and composition of cherry fruits (*Prunus avium* and *Prunus cerasus*) following fall applications of 2,4-dichlorophenoxy-acetic acid and napthaleneacetic acid. Proc. Amer. Soc. Hort. Sci. 54, 95 (1949).

— —, and B. IMHOFE: Histological changes in bindweed and sow thistle following applications of 2,4-dichlorophenoxyacetic acid in herbicidal concentrations. Bot. Gaz. 107, 62 (1945).

UOTA, M., and D. H. DEWEY: Respiration and volatile emanation of Bartlett pears as influenced by post-harvest treatment with ethylene and 2,4,5-T. Proc. Amer. Soc. Hort. Sci. 61, 257 (1953).

VAN OVERBEEK, J.: Applications of auxins in agriculture and their physiological bases. VIII. Physiology and biochemistry of 2,4-D. In: Handbuch der Pflanzen-physiologie. W. Ruhland, ed. Vol. XIV, pp. 1149-1155. Berlin: Springer-Verlag 1961 a.

— New theory on the primary mode of auxin action. In: Plant growth regulation, 4th Internat. Conf. on Plant Growth Regulation. Pp. 449-455. Ames: Iowa State Univ. Press 1961 b.

— Survey of mechanisms of herbicide action. In: Physiology and biochemistry of herbicides. L. J. Audus, ed. Pp. 387-400. New York: Academic Press 1964.

—, R. BLONDEAU, and V. HORNE: Trans-cinnamic acid as an anti-auxin. Amer. J. Bot. 38, 589 (1951).

VASCAUTANU, E., V. JURCA, and V. ARTENIE: Properties of some organic com-pounds which act on plants. Analele Stiint. Univ. "A. I. Cuza," Iasi Sect. I. 7, 193 (1961).

VELDSTRA, H.: The relation of chemical structure to biological activity in growth substances. Ann. Rev. Plant Physiol. 4, 151 (1953).

VERNON, L. P., and S. ARONOFF: Metabolism of soybean leaves. IV. Translocation from soybean leaves. Arch. Biochem. Biophys. 36, 383 (1952).

VLASYUK, P. A., and E. P. STARCHENKOV: Effect of physiologically active sub-stances on phosphorus metabolism of corn and sugar beet plants. Primenenie Mikroelementov, Polimerov and Radiaktion. Izotapov v Sel'sk. Khoz. Ukr. Akad. Sel'skokhoz. Nauk, Tr. Koordinats. Soveshch. 1960, 66 (1962).

VOROB'EV, F. K.: The selective 2,4-D inhibition of plant protein synthesis. Doklady Moskov. Sel'skokhoz. Akad. im K. A. Timiryazeva, No. 47, 117 (1959).

—, and A. A. ABUEVA: Effect of triethanolamine salt of 2,4-D on nitrogen metabol-ism in plants of flax. Doklady Moskov. Sel'skokhoz. Akad. im K. A. Timiryazeva, No. 52, 347-355 (1960).

—, and JU-PI CH'A: Effect of simazine and 2,4-D on nitrogen metabolism of plants. Doklady Moskov. Sel'skokhoz. Akad. im. K. A. Timiryazeva, No. 57, 63 (1960).

WAGENKNECHT, A. C., A. J. RIKER, T. C. ALLEN, and R. H. BURRIS: Plant-growth substances and the activity of cell-free respiratory enzymes. Amer. J. Bot. 38, 550 (1951).

WAIN, R. L.: A new approach to selective weed control. Ann. Applied Biol. 42, 151 (1955).

— The oxidative metabolism of certain auxins and their derivatives within plant tissues. In: Recent advances in botany. The growth substances and their action. Pp. 1083-1088. Toronto: Univ. of Toronto Press 1961.

—, P. P. RUTHERFORD, E. W. WESTON, and C. M. GRIFFITHS: Effects of growth regulating substances on inulin-storing plant tissues. Nature 203, 504 (1964).

WASSBERG, C., and F. J. GOODRICH: The anatomical effects produced in the leaves of *Datura stromonium* by the action of 2,4-dichlorophenoxyacetic acid. J. Amer. Pharm. Assoc. 45, 495 (1956).

WEAVER, M. L.: Factors influencing the tolerance of *Pisum sativum* to MCPA (2-methyl-4-chlorophenoxyacetic acid). Ph.D. Dissertation. Univ. of Minn., Min-neapolis (1961).

WEDDING, R. T. and M. K. BLACK: Uncoupling of phosphorylation in *Chlorella* by 2,4-dichlorophenoxyacetic acid. Plant and Soil 14, 242 (1961).

WEDDING, R. T., and M. K. BLACK: Response of oxidation and coupled phosphorylation in plant mitochondria to 2,4-dichlorophenoxyacetic acid. Plant Physiol. **37**, 364 (1962).
— — Kinetics of malic dehydrogenase inhibition by 2,4-dichlorophenoxyacetic acid. Plant Physiol. **38**, 157 (1963).
— — Interaction of nucleotides with auxins in the growth of pea stem segments. Plant Physiol. **39**, 799 (1964).
—, L. C. ERICKSON, and B. L. BRANNAMAN: Effect of 2,4-dichlorophenoxyacetic acid on photosynthesis and respiration. Plant Physiol. **29**, 64 (1954).
WEINBERGER, J. H.: Effect of 2,4,5-trichlorophenoxyacetic acid on ripening of peaches in Georgia. Proc. Amer. Soc. Hort. Sci. **57**, 115 (1951).
WEINTRAUB, R. L.: Mechanisms of action of 2,4-D. J. Agr. Food Chem. **1**, 250 (1953).
—, J. W. BROWN, M. FIELDS, and J. ROHAN: Metabolism of 2,4-dichlorophenoxyacetic acid. I. $C^{14}O_2$ production by bean plants treated with labeled 2,4-dichlorophenoxyacetic acids. Plant Physiol. **27**, 292 (1952 a).
— —, and J. N. YEATMAN: Recovery of growth regulator from plants treated with 2,4-dichlorophenoxyacetic acid. Science **111**, 493 (1950).
—, J. N. YEATMAN, J. A. LOCKHART, J. H. REINHART, and M. FIELDS: Metabolism of 2,4-dichlorophenoxyacetic acid. II. Metabolism of the side chain by bean plants. Arch. Biochem. Biophys. **60**, 277 (1952 b).
WELLER, L. E., R. W. LUEBKE, C. L. HAMNER, and H. M. SELL: Changes in chemical composition of the leaves and roots of red kidney bean plants treated with 2,4-dichlorophenoxyacetic acid. Plant Physiol. **25**, 289 (1950).
WEST, F. R., JR., and J. H. M. HENDERSON: The effect of 2,4-dichlorophenoxyacetic acid and various other substances upon the respiration of blue lupine seedling roots. Science **111**, 574 (1950).
WEST, S. H., J. B. HANSON, and J. L. KEY: Effect of 2,4-dichlorophenoxyacetic acid on the nucleic acid and protein content of seedling tissues. Weeds 8, 333 (1960).
WHITE, D. G.: Promotion of red color of apples. II. Effects of preharvest sprays of certain chemicals in multiple combinations. Proc. Amer. Soc. Hort. Sci. **61**, 180 (1953).
WHITWORTH, J. W.: Selective action of a growth regulator on strains of bindweed (*Convolvulus arvensis*). Ph.D. Dissertation. Wash. State Univ., Pullman (1961).
WILCOX, M., D. E. MORELAND, and G. C. KLINGMAN: Aryl hydroxylation of phenoxyaliphatic acids by excised roots. Physiol. Plantarum **16**, 565 (1963).
WILDON, C. E., C. L. HAMNER, and S. T. BASS: Effect of 2,4-D on the accumulation of mineral elements in tobacco plants. Plant Physiol. **32**, 243 (1957).
WILLIAMS, G., JR.: The effects of 2,4-D on mustard plants as modified by light quality. Ph.D. Dissertation. Univ. of New Hampshire, Durham (1963).
WILLIAMS, M. C., and E. H. CRONIN: Effects of silvex and 2,4,5-T on alkaloid content of tall larkspur. Weeds **11**, 317 (1963).
WITSCH, H. v., and H. KAPSERLIK: The effect of 2,4-dichlorophenoxyacetic acid upon the aneurin content of sensitive or insensitive plants. Planta **45**, 264 (1955).
WOLF, D. E., G. VERMILLION, A. WALLACE, and G. H. AHLGREN: Effect of 2,4-D on carbohydrate and nutrient-element content and on rapidity of kill of soybean plants growing at different nitrogen levels. Bot. Gaz. **112**, 188 (1950).
WOODBRIDGE, C. G.: The effect of some insecticides and 2,4-D on the sugar content of Bartlett pear tissues. Proc. Amer. Soc. Hort. Sci. **81**, 123 (1962).
—, and A. L. KAMAL: The effect of 2,4-D on the nitrogen fractions of Bartlett pear tissues. Proc. Amer. Soc. Hort. Sci. **81**, 116 (1962).
WOODFORD, E. K., K. HOLLY, and C. C. McCREADY: Herbicides. Ann. Rev. Plant Physiol. **9**, 311 (1958).
WOOFTER, H. D.: Retention and effect of 2,4-D sprays on winter wheat. Ph.D. Dissertation. Ohio State Univ., Columbus (1959).

WORT, D. J.: The response of buckwheat to treatment with 2,4-dichlorophenoxy-acetic acid. Amer. J. Bot. 36, 673 (1949).
— Effects of non-lethal concentrations of 2,4-D on buckwheat. Plant Physiol. 26, 50 (1951).
— Influence of 2,4-D on enzyme systems. Weeds 3, 131 (1954).
— Effects on the composition and metabolism of the entire plant. In: Handbuch der Pflanzenphysiologie. W. Ruhland, ed. Vol. XIV, pp. 1110-1136. Berlin: Springer-Verlag 1961.
— Effects of herbicides on plant composition and metabolism. In: Physiology and biochemistry of herbicides. L. J. Audus, ed. Pp. 291-334. New York: Academic Press 1964 a.
— Responses of plants to sublethal concentrations of 2,4-D, without and with added minerals. In: Physiology and biochemistry of herbicides. L. J. Audus, ed. Pp. 335-342. New York: Academic Press 1964 b.
—, and L. M. COWIE: Effect of 2,4-dichlorophenoxyacetic acid on phosphorylase, phosphotase, amylase, catalase, and peroxidase activity in wheat. Plant Physiol. 28, 135 (1953).
WORTH, W. A., JR., and A. M. McCABE: Differential effects of 2,4-D on aerobic, anaerobic, and facultative anaerobic microorganisms. Science 108, 16 (1948).
WRIGHT, D. E.: Effect of plant growth regulators on uptake of amino acids by plant roots. Nature 192, 1044 (1961).
YAKUSHKINA, N. I.: Effect of growth stimulators on distribution of nutrients in plants. Doklady Akad. Nauk. S.S.S.R. 69, 101 (1949).
— Effect of growth stimulants on phosphorus metabolism in tomatoes. Doklady Akad. Nauk S.S.S.R. 109, 635 (1956).
—, and B. E. KRAVTSOVA: The influence of growth stimulants on the yield and quality of the fruit of several vegetables. Doklady Vsesoyuz. Akad. Sel'skokhoz. Nauk im. V. I. Lenina 22, 15 (1957).
—, and T. V. LIKHOLAT: Effect of 2,4-D on oxidative phosphorylation in plants of various systematic groups. Khim. v Sel'sk. Khoz. 12, 56 (1964).
YAMADA, N., and Y. MURATA: The effect of 2,4-D on wheat seedlings. Nôgyô Oyobi Engei 25, 169 (1950).
YASUDA, G. K., M. G. PAYNE, and J. L. FULTS: Effects of 2,4-dichlorophenoxy-acetic acid and maleic hydrazide on potato proteins as shown by paper electrophoresis. Nature 176, 1029 (1955).
ZELITCH, I.: Biochemical control of stomatal opening in leaves. Proc. Nat. Acad. Sci. U.S. 47, 1423 (1961).
ZEMSKAYA, V. A., and Y. V. RAKITIN: Detoxification of 2,4-D in sunflower and oat plants. Agrokhimiya 7, 101 (1964).
ZENK, M. H.: Enzymatische Aktivierung von Auxinen und ihre Konjugierung mit Glycin. Z. Naturforsch. 15b, 436 (1960).
ZIMMERMAN, P. W.: Present status of plant hormones. Ind. Eng. Chem. 35, 596 (1943).
ZOSCHKE, M.: The effect of synthetic hormone-herbicides on crops and weeds. Kuhn-Arch. 71, 305 (1957).
ZWAR, J. A., and A. H. G. C. RIJVEN: Inhibition of transport of 3-indoleacetic acid in the etiolated hypocotyl of Phaseolus vulgaris. Australian J. Biol. Sci. 9, 528 (1956).

Review of the symposium on foreign materials in food, Lucerne, April 8 to 9, 1965*

By

S. Dormal-van den Bruel**

Contents

I. Introduction 114
II. Fundamental considerations about the problem of foreign
materials in foodstuffs 115
III. Problems relating to intentional food additives 116
IV. Problems relating to non-intentional food additives 118
V. Food additives in relation to consumer protection,
public health, and food control in Switzerland 123
VI. Conclusions and comments 125
Summary 125
Résumé 125
Zusammenfassung 126

I. Introduction

An international symposium "Foreign Materials in Food" was held in Lucerne, Switzerland, on April 8 and 9, 1965. It was organized by Professors Drs. H. AEBI (Bern), E. GRANDJEAN (Zürich), and O. HÖGL (Bern), under the auspices of the Swiss Academy of Medical Sciences, the Swiss Society of Research on Nutrition, the Swiss Society of Preventive Medicine, and the Federal Commission for Food, Legislation, and Food Control. Distinguished Swiss and foreign specialists in the matter of food additives were invited to take the floor. Their reports were attended by a large audience and formed the subject of numerous interventions, where various trends and opinions with regard to the admissibility of foreign matters in foodstuffs were confronted.

The session was opened by Dr. A. SAUTER, Director, Federal Department of Public Health, Bern, and the aim of the symposium was presented by

---

* Owing to the interest of the subject, the Editorial Board of RESIDUE REVIEWS considered it worthwhile to review in summary form this entire symposium, although most of the individual reports were published *in extenso* in a special issue of the ZEITSCHRIFT für PRÄVENTIV MEDIZIN, vol. 10, part 4, July-August 1965.

** Communauté Economique Européenne, Commission, Direction Générale de l'Agriculture, 12, av. de Broqueville, Bruxelles 15.

Professor H. AEBI, who gave a general survey of the problem of foreign matters in food. The admixture of food additives to foodstuffs continuously sets new problems before the authorities who assume the responsibility of public health. The data which, at first sight, may appear clear and simple often show, when carefully scrutinized, much complexity and become matters of serious debates when decisions are to be reached. Various trends are asserted, namely as to the limits and restrictions to impose in the admissibility of these compounds and to what extent food has to be kept under control. The answer to these problems requires a conscientious investigation of the situation based on all the available scientific knowledge.

In organizing this meeting, the committee felt that a discussion among specialists could contribute to clear up some problems of general interest and to help the public in acquiring rational knowledge on this complex matter.

II. Fundamental considerations about the problem of foreign materials in foodstuffs

This subject was discussed by Professor O. HÖGL (Bern). According to him, the reserved or hostile attitude of the consumer towards food additives arises either from the fear of some hazard or from the respect for ancestral food habits, and is related to a lack of knowledge on the subject. It must be borne in mind that the evolution of the way of living and, namely, the continuously increasing population and the reduction of the working time impose modifications in food production and processing techniques. In order to increase yields and to allow transportation and storage of foodstuffs, it is necessary to have recourse to mechanization and to artificial means that unavoidably result in modifications in the composition of food. The application of these techniques is a matter of extensive research in numerous laboratories, and of responsibility for public health authorities.

As far as *intentional additives*—such as food colors, chemical preservatives, or emulsifiers—are concerned, measures were already taken as far back as 1870 in most of the developed countries for assuming protection of the consumer against hazards of intoxication. At that time, the basic principle of safety was the prohibition of the use of any compound known to involve acute toxicity hazards. Later on, it was recognized that chronic toxicity had to be taken into consideration, and that lack of data on toxic effects was not an evidence of harmlessness.

At present, and already for more than ten years, the principle of positive lists has been accepted by most governments in Europe and North America. This implies that the approval of any new food additive requires the submission of data showing the evidence of harmlessness on the basis of short- and long-term toxicity experiments on several animal species, and that compounds that are not mentioned on permitted lists are prohibited.

The protection of the consumer against *non intentional additives*, such as pesticide residues, is based on the same principle. However, since pesticides need often to be used in amounts which may be toxic for mammals, recommendations are made as regards to the elapsed time between last treatment of the crop and harvest, in order to allow the residues to metabolize or disappear. Furthermore, limits or tolerances for pesticide residues have been established and enforced by law in many countries.

In closing, Professor HÖGL felt that the problem of foreign matters in food does not appear as important for Switzerland as for some other countries, since serious restrictions in the use of these compounds were already set up in 1926 in this country. However, some questions, such as the declaration of chemical additives on foodstuffs at wholesale and retail outlets, and the extent to which a compound ought to be considered as "foreign matter" in food needs to be cleared up.

III. Problems relating to intentional food additives

This subject is limited to the three following groups of food additives: *food colors, chemical preservatives,* and *emulsifiers* and *stabilizers.*

Professor R. TRUHAUT (Paris) dealt with *coloring matters in foodstuffs.* The admixture of food colors to foodstuffs requires the establishment of *criteria of purity, conditions of use,* and *safety evaluation.* Criteria of purety need to be established for each food color in order to avoid the presence of impurities that may be noxious. Conditions of use must be rigorously delimited for avoiding chemical modifications that might arise from the reaction of the food color with some food components. Safety evaluation must be based on short- and long-term toxicity experiments on several animal species, for the purpose of detecting any possible harmful effect and, in particular, carcinogenic effects that might be involved either by the coloring matter itself or by its metabolites. The lecturer showed with many examples that, under the influence of some ferments or enzymes in the body, harmless food colors may be reduced or metabolized into compounds which may induce toxicity or carcinogenesis. Furthermore, natural coloring matters must be submitted to the same criteria as synthetic ones. In opposition to a common public opinion, "it is in nature that the most drastic poisons are found," Professor TRUHAUT mentioned as examples, hematoxylin, Persian yellow, and other vegetable and animal coloring matters.

Among food colors that were submitted to extensive toxicological studies, only about ten were, as of this date, evaluated as quite safe. Others are very likely harmless, but their toxicological data are not yet comprehensive enough for a definite conclusion.

The storage of foodstuffs by means of chemical preservatives was discussed by Professor W. DIEMAIR (Frankfort). Chemical additives used to prolong storage of foodstuffs are compounds that have bactericidal, fungicidal, bacteriostatic, or fungistatic properties which allow blocking or inhibiting of biochemical cell processes.

These compounds may also impair the color, the taste, or the flavor of food, or adulterate some of their components, such as albumins, lipids, enzymes, and vitamins. In high amounts, they may have some toxic action on the human body and be stored in various organs and tissues. Harmful effects may also arise from the chronic absorption of low levels of additives having cumulative or co-carcinogenic properties. Furthermore, some antigen action and a decrease in the resorption of "essential substances" (essentieller Stoffe), with influences on the intestinal flora and inhibition of the digestion ferments were reported. Professor DIEMAIR gave a detailed account of these phenomena on the basis of numerous papers published in the literature and of his own research. Finally, the properties of the following food preservatives were examined: biphenyl and orthophenylphenol, sorbic acid, the diethyl ester of pyrocarbonic acid, and sulfurous acid.

Biphenyl and *orthophenylphenol* are used for the protection of citrus fruits against molds of the *Diploidia* species. Owing to its high solubility in waxes and essential oils, biphenyl is stored in the rind of the fruits and only small amounts of the compound penetrate into the pulp and juice. The amount of biphenyl that is normally ingested by the consumption of peeled fruits does not imply any hazard for human health. This compound is not stored in the body.

Sorbic acid is used in many countries for preserving foodstuffs. It is found in the form of the lactone in sorb fruits. It has the property of inhibiting essential fermentive processes; however, as with many other natural fatty acids, it decomposes in the body into carbonic acid and water. Therefore, it does not involve any hazard.

Diethyl ester of pyrocarbonic acid (PKDE) has been recently proposed for the preservation of wine, beer, and natural and fermented fruit juices. At dosages of 200 to 800 mg./1. it has an efficient microbiocidal action on yeasts and inhibits fermentation during a few hours after its admixture. In the presence of water, it decomposes into ethyl alcohol and carbonic acid with formation of small amounts of diethyl carbonate which are proportional to the percentage of alcohol present in the liquor. Diethyl carbonate adulterates proteins and vitamin C; therefore, the use of PKDE is not allowed for the preservation of beverages in Germany. However, recent physiological research tends to prove that tolerances of one p.p.m. for PKDE and ten p.p.m. for diethyl carbonate in beverages are admissible.

Sulfurous acid is frequently used as a sterilizing agent. It has been shown that its toxicity varies considerably according to individual sensitivity. Amounts as low as ten mg. have produced noxious effects, while, in other cases, 2,000 mg. were tolerated perfectly. The mode of action of sulfurous acid was discussed in detail, and a scheme as to the possible fates of the sulfites in the body was proposed.

Professor J. W. REITH (Utrecht) then dealt with the problem of *emulsifiers and emulsion stabilizers in foodstuffs*. Comparative lists of emulsifiers and stabilizers authorized in the United States, the United Kingdom, and the member countries of the EEC were presented. On the whole, they include about 70 natural and synthetic compounds belonging to various chemical groups and sharing the property of physically binding two immiscible phases.

The fundamental problem set by these compounds is to know whether they are undesirable in foodstuffs by reason of their possible influences on the resorption of some substances by the intestines. This question was carefully discussed on the basis of numerous experimental studies. Although resorption by the intestines of some toxic compounds, coloring matters, lipids, vitamins, and cholesterol was shown to be promoted by admixture of high amounts of emulsifiers, harmful effects could not be demonstrated in the case of normal diets provided with low levels of surfactants such as those that are commonly used in food technology. Therefore, Professor REITH was of the opinion that there is no reason to prohibit the use of emulsifiers or stabilizers when they meet the criteria that are imposed.

In 1964, the joint FAO/WHO experts committee established maximum acceptable daily intakes for about twenty emulsifiers and stabilizers. In doing this work, the committee attached an essential importance to the properties of enzymatic hydrolysis, resorption, metabolism and excretion of the compounds under consideration. In general, toxicological evaluation was favorable when it was biochemically shown that hydrolysis was fast and total and that degradation compounds were not resorbed.

The number of emulsifiers and stabilizers that are authorized appears to be much higher in the United States and in the United Kingdom than in Western Europe. The criteria governing the admission of these compounds are not known for every country. However, some reticence towards the use of synthetic materials seems to occur in Europe. As already outlined by Professor TRUHAUT with regard to coloring matters, Professor REITH considered that making a distinction between natural and synthetic matters is erroneous. Only results of long-term toxicity experiments may bring evidence of the harmlessness of a compound.

An increase in the use of synthetic emulsifiers and stabilizers can be expected only if industry submits the results of its research to governmental authorities who are responsible for the approval of food additives. Among other data, the total amount of an emulsifier or stabilizer that may be ingested with each foodstuff as well as the total daily food intake needs to be taken into consideration in evaluating the maximum permissible level. This principle has already been applied by the joint FAO/WHO experts committee (1964) with regard to phosphates.

IV. Problems relating to non-intentional food additives

The two following groups of non intentional food additives were examined: *substances migrating from packagings* and *pesticide residues*.

Professor R. FRANCK (Berlin) dealt with *artificial materials as packaging materials for foodstuffs*. The most frequently used materials for food packaging are either transformation products from natural origin, such as parchment, cellulose acetate, rubber, vulcanized fiber, linoleum, galalith, etc., or synthetic materials, such as polyvinyl chloride, polyethylene, polystyrene, polyamide, polyacrylate, etc. From the chemical standpoint, these compounds are macromolecules ranging about 1,000 to 10,000 times the size of usual organic molecules. They are obtained by polymerization, polycondensation, or polyaddition of single molecules, called monomers.

From the health standpoint, macromolecules do not involve any hazard since they are insoluble in water and most solvents, and, with a few exceptions, stable towards food acids, oils, fats, and alcohols. However, in practice, macromolecules are never pure. They always contain residues of monomers or intermediary compounds of low molecular weight and auxiliary substances, such as catalysts, stabilizers, emulsifiers, protective colloids, antistatics, and filling and softening materials which are used in the manufacturing processes. These compounds are able to migrate from the packaging material into foodstuffs and may also have a "carrying" action in helping certain compounds to migrate into the artificial material. Therefore, the toxicological evaluation of artificial materials for food packaging is extremely difficult.

Several national and international organizations are studying this problem, and regulations have already been elaborated in some countries. Professor FRANCK discussed the basic principles to be adopted for protecting the consumer against health hazards which might arise from the absorption of substances migrating from packagings into food. The establishment of positive lists of permitted materials with prescriptions as to the processes of investigation of the materials and the packaged food appears, in the present state of our knowledge, to be satisfactory. The legal proceedings used in the U.S.A. and the U.K. were briefly reviewed. In Western Germany, the 1958 Food Law prohibits the use of tools and, in particular, of packaging materials, under conditions where technically unavoidable foreign materials that are not harmless or that may impair the flavor or the taste of food could migrate into foodstuffs. For each artificial material and each use data about harmlessness and chemical composition are to be provided by *the user*. However, in order to relieve him of his responsibility, recommendations as to the evaluation processes are elaborated by the "Kommission für die gesundheitliche Beurteilung von Kunststoffen und anderen Polymeren," and lists of permitted substances are published by the Department of Public Health. The basic principles that have been adopted in the recommendations are the following:

(a) Food must respond to the general criteria of purity imposed by the German Food Laws. The control of foodstuffs as to the presence of artificial materials of low molecular weight has to be performed according to special tests that are described.

(b) The evidence of harmlessness about auxiliary substances in the packaging materials ready for use must be kept according to methods internationally adopted. The toxicological evaluation may, however, be restricted in each case to the proposed uses of the material under consideration.

(c) The compounds whose presence is admitted in foodstuffs are quantitatively limited, and their methods of analysis are described.

The problem of *pesticide residues in foodstuffs* was discussed from its analytical standpoint by Dr. S. DORMAL-VAN DEN BRUEL (Brussels) and from its biological standpoint by Professor H. MAIERBODE (Bonn).

Dr. S. DORMAL-VAN DEN BRUEL stated that numerous pesticides that are used for the treatment of plants and stored foods leave on the subtrates residues that may penetrate into cuticle, epidermis, and adjacent tissues or dissolve in fats and oils. These residues may persist as such or be metabolized during periods ranging from a few hours to several months after treatment. Similar phenomena occur in root-plants following soil treatments with persistent pesticides. Moreover, ingestion by cattle and poultry of treated feeds may result in the presence of some residues in animal fats, meat, milk and milk products, and eggs.

In many countries measures have been taken to protect the consumer from health hazards that might arise from the ingestion of pesticide residues. According to the principle established by the joint FAO/WHO experts committee (1961), maximum amounts of residues that are permitted in foods are those which result from "good agricultural practices" providing that they are not higher than the toxicological limits evaluated as safe for long consumption by man. These conditions require the determination of the *persistence* and *metabolism* of each pesticide on every treated crop and its by-products, the *natures* and *amounts* of the residues at the time of harvest or slaughter, their *acute and chronic toxicities* on several animal species, and their possible effects on man. When systems of limits or tolerances are enforced by decree or law, a control of pesticide residues in foods at wholesale and retail outlets must be achieved. This implies that analytical procedures be set up in order to detect, isolate, identify, and determine amounts of residues as low as a few micrograms in large bulks of organic materials. However, the methods and techniques to apply depend on the particular problems to be solved.

The most currently used procedures of analysis were discussed in relation to the various problems posed to the residue analyst, and the importance of the criteria of *specificity*, *accuracy*, and *sensivity* for the choice of a method were outlined. Special attention was drawn to the difficulties involved by the *detection* and *identification* of pesticide residues in samples of unknown history, such as those usually submitted to official laboratories dealing with control of foods offered for consumption. In connection with this aspect of the subject, the possibilities and limits of applicability of *multi-detection techniques*, such as paper, thin-layer, and gas chromatography, and electrophoresis were discussed, and collaborative studies undertaken by international organizations such as AOAC, EPPO, the British Analytical Panel[1], and EEC were briefly reviewed.

---

[1] The Analytical Panel of the Scientific Sub-Committee of the Advisory Committee on Poisonous Substances used in Agriculture and Food Storage.

To close, Dr. DORMAL-VAN DEN BRUEL surveyed the amounts and natures of pesticide residues found in foodstuffs offered for consumption in
the United States, Canada, The Netherlands, Switzerland, and the United
Kingdom. It comes out that, in most cases, residue levels are much lower
than the permitted limits or tolerances. However, residues of some organochlorine compounds, such as DDT, DDE, lindane, and dieldrin are frequently
encountered in any kind of food and, in particular, in staple foods such as
milk and milk-products. This correlates with the occurrence of these compounds in human adipose tissues, as demonstrated by different scientists in
the U.S.A., the U.K., Western Germany, France, and Israel. Although the
presence of these residues in the human body has not resulted in health injuries, the lecturer was of the opinion that a discrete attitude as regards the
use of pesticides and frequent controls of foodstuffs offered in markets are
desirable.

Professor H. MAIER-BODE then showed the importance of determining
the amounts of pesticide residues that may persist on treated plants, since
the effects of toxic compounds are largely dependent on the quantities that
are ingested. Methods of calculation of initial deposits, persistence curves,
and requisite time between last treatment of the crop and harvest, in order
to allow the residue values to drop down to maximum acceptable intakes for
man, were presented and illustrated by several examples.

As far as acute toxicity, based on LD_{50} values, is concerned, the most
toxic pesticides are essentially insecticides from the group of organophosphorus compounds. However, the use of these pesticides does not involve
health hazards for the consumer since they generally metabolize or decompose into harmless compounds within a short time after application. On the
other hand, the acute toxicity of the less toxic pesticides is of the same order
of magnitude as that of numerous spices, such as vanillin, menthol, etc.,
which are currently used in rather important amounts in food. Still the main
problem in relation with the consumer's health is to know which are the effects on the organism of the daily ingestion of persistent pesticide residues
that have cumulative properties. In the United States, average amounts of
ten p.p.m. of DDT and DDE, 0.6 p.p.m. of lindane, and 0.1 p.p.m. of dieldrin
were found in human adipose tissues. Research conducted by the lecturer on
60 inhabitants in Western Germany showed that the average amount of DDT
and DDE in adipose tissues was 2.3 p.p.m., the highest amounts occurring
in older subjects. According to experimental results on animal species, the
storage of DDT in adipose tissues seems to protect the organism against the
noxious effects of this compound. On the other hand, the possible metabolism
of DDT deposits in the organism has been proved inconsequent, owing to
the small amount of toxic material that is involved. Nevertheless, it appears
undesirable that significant amounts of pesticide residues be stored in the
human body. Therefore, the use of persistent pesticides is as much as possible
restricted. Permissible levels that may be ingested by man during his life-
time without producing any toxic action have been evaluated for a certain

number of pesticides by a joint FAO/WHO experts committee. From these data, maximum acceptable daily intakes for pesticide residues in foodstuffs may be calculated, and restrictions in the use of some pesticides, in order to meet these requirements, established. In fact, most of the herbicides and a certain number of insecticides and fungicides totally decompose within the time elapsed between last application of the pesticide and harvest of the crop.

It has also been asked whether pesticides do not exercise any influence on some components of agricultural commodities, such as sugars, amino-acids, vitamins, etc., and on the soil microflora. As of this date, no deleterious effect has been shown.

These reports were followed by a long discussion where Drs. H. L. HALLER (Washington), H. MÜLLER (Grosshöchstetten), and H. GYSIN (Basel), Mrs. P. MAAG (Zürich), Dr. F. SCHNEIDER (Wädenswil), and Professor W. SCHUPHAN (Geisenheim am Rhein) successively expressed their opinions on pesticide residues. Dr. HALLER reported the proceedings followed in the United States in order to protect the consumers' health against hazards that might arise from the use of pesticides. Dr. GYSIN stated the responsibilities of industry in that matter, while Dr. SCHNEIDER showed the sustained efforts that are accomplished in Switzerland by the federal research stations and the cantonal and municipal laboratories in collaboration with university and industrial laboratories to insure a rational use of pesticides and the consumers' safety. After Mrs. MAAG's contribution, which reflected the opinion of an optimistic journalist, arguments against the use of pesticides were developed by Dr. MÜLLER and Professor SCHUPHAN.

Dr. MÜLLER, who is a defender of "organic farming," showed that the use of chemicals is not necessary to obtain fruits and vegetables free from pests.

Professor SCHUPHAN, who is a specialist in nutrition and food quality, pointed out the disadvantageous effects on food quality involved by the use of pesticides or food additives, such as the treatment of citrus fruits with *biphenyl* and *orthophenylphenol*, and the coating of these fruits with waxes. Even in very low amounts, biphenyl and orthophenylphenol impair the taste and smell of the fruits. Moreover, although these compounds allow the prolongation of the normal time of storage of the fruit and keep it in perfect appearance, a decay in the food value and, namely, a decreasing vitamin C content cannot be avoided. The commercialization of such treated fruits thus constitutes a fraud towards the consumer and a breach of the food regulations which require food additives to be free from flavor. Also, it is not unlikely that some allergies or mucous membrane irritations by fruit wholesalers is caused by the presence of biphenyl and orthophenylphenol residues in the rinds. Therefore, the lecturer was of the opinion that declaration of the treatments, as prescribed in Germany, needs to be done. On the other hand, the use of biphenyl and orthophenylphenol appears to be useful only if the consumer cares to eat citrus fruits throughout the year.

V. Food additives in relation to consumer protection, public health, and food control in Switzerland

The last part of the symposium was devoted to the following subjects: *food additives and the consumer's protection* by Professor O. ANGEHRN (Zürich), *food additives and public health* by Dr. O. JEANNERET (Geneva), and *possibilities and limits of food control on the basis of Swiss food regulations* by Dr. E. MATTHEY (Lausanne).

According to Professor ANGEHRN, the consumers may be grouped into three categories: those who consider that the market economy is unfavorable to the consumer and require a drastic intervention from the State, those who critically examine the situation and feel they can themselves contribute to improve it, and finally those—the most numerous ones—who have no opinion or only hazy knowledge on the subject. In connection with this aspect, a committee has been set up by the Swiss Parliament in order to examine the consumers' interests.

In general, it is recognized that free market economy is the more satisfactory as a certain equilibrium is shared by sellers and purchasers. This means that, from both sides and also by the consumer, knowledge, information, and a reasonable attitude prevail. It is obvious that the State has the responsibility to use its technical and judicial means to preserve the consumer from any health injury due to food consumption, and much has been done in this regard by the Swiss State; however, an effort should be made by the consumer himself who has to get accustomed to "live with chemistry."

It is the consumer's privilege to request appropriate declaration of food additives on food offered in markets, whatever the nature of the additives may be. Whether the declaration is helpful or not to the consumer, who generally lacks requisite knowledges to appreciate the real value of a chemical treatment, has not to be taken into consideration. It appears indeed reasonable and desirable that the consumer wants to know exactly what he is purchasing and eating. A particular case that undoubtedly needs declaration is the one of beverages that are treated with stimulating agents that are harmless but not necessarily free from any effect on some subjects. Furthermore, the consumer must be in a position critically to examine advertisements. In conclusion, Professor ANGEHRN was of the opinion that the consumer must be informed and educated in order to play his part in the market economy and to become conscientious of his interests.

Dr. JEANNERET then discussed the responsibilities of public health agencies in the matter of food additives. In this field, the concern to keep population in good health has first brought up national legislation, the basis and trends of which are sometimes divergent. At the international level, general principles are later developed and adopted. However, their practical application often comes up against methodological or technical difficulties, namely, when toxicologists and epidemiologists have to assess long-term toxicity hazards.

Although already initiated, an international cooperation appears necessary for political, economic, and scientific reasons. It should be based on collaborative studies performed by groups of scientists including chemists, biochemists, physicists, nutrition specialists, food technologists, pathologists, pharmacologists, and toxicologists.

On the other hand, food control services of public health agencies should be supplemented by information and education services for helping the consumer progressively to acquire more rational knowledge and to adopt a conscientious attitude towards food additives. This could be undertaken at the secondary school level.

Dr. MATTHEY recalled the principles of the Swiss legislation regarding food control. The Swiss Food Law is a federal law dating back to 1905, conferring to the cantons the responsibility to enforce it. Its purpose was to protect the consumer against injuries that might arise from food deterioration or alteration, or improper manufacture. However, owing to its adaptable judicial structure, it still applies to the use of modern food additives. The admixture of food additives to foods is prohibited unless actual provisions are prescribed. This principle allows most of the foodstuffs to be free from food additives, and to establish positive lists of permitted additives whose use is restricted to certain categories of foods and beverages. As far as pesticide residues are concerned, the Food Law prescriptions are not adapted to the present situation, although they implicitly allow the presence of trace amounts of pesticide residues in agricultural commodities. This chapter needs to be revised and very likely the establishment of limits or tolerances will be faced.

On the other hand, the control of food additives and pesticide residues in foods offered for consumption, the responsibility of which lies with cantonal laboratories, at present raises serious difficulties. Owing to the chemical complexity of some compounds permitted as food additives and to the fact that pesticide residues are generally present in trace amounts in foods, analytical procedures require the use of modern instrumentation, often very burdensome and rapidly becoming obsolete. Moreover, they need, in most cases, tedious researches which only can be performed in laboratories provided with a highly specialized scientific staff. These difficulties result in a restriction in the food control in Switzerland, mainly because of the federal structure of the State.

In conclusion, DR. MATTHEY thought that the basic principles of the Swiss Food Law need not be modified. However, in order to improve the situation as regards food control, it should be desirable that analytical research or at least coordinated work in that field be undertaken by federal laboratories—such as is already done in agriculture. Furthermore, a judicious distribution within cantonal laboratories of control analyses requiring special or burdensome instruments should be helpful.

VI. Conclusions and comments

The conclusions of the symposium were drawn by Professor H. AEBI following a general discussion contributed by Dr. A. SAUTER, Director, Swiss Federal Department of Public Health.

The essential object of the discussion concerned food additives in relation to the consumer's standpoint. It may be summarized as follows. A trend for an enlargement and improvement of food control in order that even minor frauds can be detected was asserted. Furthermore, compulsory declaration of food additives on foods offered for consumption was steadily demanded. The Swiss consumer wants to know exactly what he is purchasing and eating, although it appears he often lacks scientific knowledge to appreciate the reason for a treatment or to clear up the nature of a chemical.

The rapporteur should like to emphasize the interest of this symposium and, in particular, the quality of its scientific contributions. As regards conclusions, one might, however, regret that more attention was not drawn to some scientific problems of special interest, such as the one of plastic materials for food packaging which use is being considerably extended, in spite of the fact that very little is known about the substances migrating from the plastics into foods. Also, the importance granted to compulsory declaration of food additives appeared to the rapporteur to be overestimated. The scope of such a measure seems indeed rather questionable in so far as food legislation is not enforced by a system of limits or tolerances and as control laboratories are not adequately equipped to perform detection, identification, and determination of any permitted or non-permitted intentional or non-intentional food additives.

Summary

The Symposium on Foreign Materials in Foods, held in Lucerne, April 8 and 9, 1965 is reviewed with a survey of the reports that were presented. The subjects that were dealt with included fundamental considerations about the problem of foreign materials in foodstuffs, problems relating to intentional and non-intentional food additives, and considerations of food additives in relation to consumer protection, public health, and food control in Switzerland.

The conclusions of the symposium are reported with some personal comments by the author.

Résumé

Le présent article est un compte-rendu du *Symposium sur les Matières étrangères dans les Aliments* qui s'est tenu à Lucerne, les 8 et 9 avril 1965. Il comporte un résumé des différents rapports qui y ont été présentés. Les sujets suivants ont été traités: considérations fondamentales sur de problème des matières étrangères dans les aliments, problèmes relatifs aux additifs intentionnels et non intentionnels, considérations sur le problème des additifs alimentaires en relation avec la protection du consommateur, la Santé Publique et le contrôle des aliments en Suisse.

Les conclusions du Symposium sont rapportées avec un bref commentaire personnel de l'auteur.

Zusammenfassung

Vorliegender Artikel ist ein Bericht betreffend das *Symposium über Fremdstoffe in Lebensmitteln,* das am 8. und 9. April 1965 in Luzern stattgefunden hat. Er enthält eine Zusammenfassung der verschiedenen dort vorgetragenen Berichte. Folgende Themen wurden behandelt: Grundsätzliche Betrachtungen zum Problem der fremden Beimischungen in Lebensmitteln, Probleme betreffend beabsichtigte und nicht beabsichtigte Zusätze sowie Betrachtungen über das Problem der Lebensmittelzusatzstoffe in Zusammenhang mit dem Verbraucher, der Volksgesundheit und der Lebensmittelkontrolle in der Schweiz.

Die Schlussfolgerungen des Symposiums werden mit einem persönlichen Kurzkommentar des Verfassers wiedergegeben.

Subject Index

Acetone, purification 4
Acetonitrile, purification 4—6, 8
Actinomycin D 53, 67, 68
Acute toxicity 115, 120
ADP 40 ff.
Agropyron pauciflorum 57
Aldrin 4, 14, 16
Alfalfa 47
Alkaloids 55
Alumina 2, 8
Ammonia 21
AMP 40 ff.
Anacharis 79
Analyst shortage 17—18
Aneurin 56
Antigen action 117
Antimycin A 77
Antistatica 119
Apples 57, 64, 88, 89
Apricots 41, 56, 63, 64, 84
Aramite 21
Arsenates 16
Artichokes 57, 59, 63, 69, 70
ATP 40 ff.
AutoAnalyzer applications 23
Automated pesticide residue analysis 12 ff.
Auxins 53—55, 59, 64, 66—70, 72, 73, 78—87
Avena coleoptile 66, 82
Ayapin 56
Azaquanine 67

Bananas 41, 55, 71
Barbarea verna 57
Barley 42, 49, 50, 54, 59, 62, 63, 91
Beans (see also specific kinds) 42, 47—50, 52, 54, 58—65, 72, 73, 76, 79—88, 90—92
Beer 117
Beets (see also Sugar beets) 41, 49, 61, 71
Benzene, purification 1, 4, 5
Benzenehexachloride 22
Beverages 123, 124
Bindweed 42, 43, 45, 46, 52, 73, 84

Biphenyl 117, 122
— AutoAnalysis 26—34
Birdsfoot trefoil 50
Brassica campestris 60
Broccoli 58
Bryophyllum calycinum 63
Buckwheat 45, 49, 50

Cabbage 78
Captan 16
Carbamates 15, 21, 23
Carbaryl 16
Carbohydrates 40 ff.
— metabolism 71
Carbon tetrachloride, purification 4, 9
Carotene 57
Carrots 41, 48
Castor beans 64, 72, 73, 88
Cats 19
Cattle 46
Celite 4
Cell elongation 81
Cellulose 42
— acetate 119
Chenopodium album 63
Cherries 41, 47, 63
Chicory 63
Chlorbenside 16
Chlordan 7, 16
Chlorella pyrenoidosa 78, 80, 81
Chlorine 21
Chlorobenzilate 8
Chlorofenson 16
Chlorofluorophenoxyacetic acid 88
Chloroform, purification 8, 9
Chlorophenoxy herbicides 39 ff.
Chlorophyll 57, 81
Cholinesterase 17, 18, 22 ff.
— activity, screening 18 ff.
Chromatography, automated 20
— for solvent purification 8
Chronic toxicity 115, 120
Cirsium arvense 43
Citrus fruits 117, 122
— rinds 28 ff.

Cleanup 19 ff.
Clover 47, 50
Cocklebur 62
Coenzyme A 40 ff., 72, 91
Colorimeters, automated 22
Combustion, automated 22
— methods 21 ff.
Commelina spp. 62
Conditioned reflexes 19
Consumer categories 123
— protection (Switzerland) 123
Copper (see also Minerals) 21
— fungicides (see also specific names)
    15, 16
Corn 42, 47, 50, 58, 60, 61, 64, 65, 68,
    71—77, 86, 88, 91
Cotton 42, 51, 53, 55, 68, 72, 84, 86,
    89, 90
— seed oil 44
2-CPA effect on plants 54
4-CPA effect on plants 40, 81, 90, 91
Crop items 12
Cucumbers 47, 51, 54, 66, 68, 71, 72,
    89, 90
Currants 89
p-Cymene 28

2,4-D effect on plants 39—92
— inhibitors (see also specific com-
    pounds) 77
— metabolites 89 ff.
— site of action 67 ff.
— translocation 81, 84, 92
Dandelions 46
Datura stramonium 54, 55
2,4-DB effect on plants 39, 65, 73, 87
DDE 121, 122
DDT 12, 15, 16, 121, 122
Deionized water, interferences in 2
Delnav, see Dioxathion
Detergents, contamination of solvents by
    2, 8
Dextrins 40, 42, 43
Dialysis, automated 20
Diazinon 16
Dibrom 15
Dichlorophenoxyisobutyric acid 70
Dicofol 16, 21
Dieldrin 6, 13, 14, 16, 121
Diethyl carbonate 117
Dilan 6
Dimethoate 4
Dimethylformamide, purification 8
Dinitrobenzene 22
Dinitrocaprylphenylcrotonate 8

Dioxathion 4, 16
Diphenylamine 6, 8, 21
Distillation of solvents 2 ff.
DNA 40 ff.
DNP effect on plants 62, 75, 77, 78
Dodine 16
2,4-DP effect on plants 65
Drift 13
Dye-body formation, automated 20

Eggplant 41, 56
Eggs 120
Electron-affinity detectors 2 ff.
Electrophoresis 22, 120
Emulsifiers 115—119
Endosulfan 16
Endrin 6, 13
Enzymes and herbicides 69—75
Ethion 16
Ethyl alcohol, purification 4
— ether, purification 6
Ethylene 55
Euonymus japonica 51, 75
Evacuated Separator 30, 31

Fat, animal 6, 120
— human 121
Fentin 15, 16, 21
Ferbam 16
Fescue 50
Filtration, automated 20
Flame photometers, automated 22
Flax 44, 49, 50
Florisil 2, 8
Fluorometers, automated 22
Folpet 16
Food colors 115, 116, 118
— carcinogenicity 116
Food control (Switzerland) 123
Foreign materials in food, symposium
    115 ff.
Foxglove 56
Fruit juices 117

Galalith 119
Galium aparine 89
Gas chromatography 2 ff., 120
— automated 20
Giberellic acid and gibberellins 56, 88
Gladiolus 66
Glutathione 69
Grains (see also specific kinds) 58
Grapefruit 28
Grasses 47, 50, 70, 72

Halogen screening 18 ff.
Helianthus annuus 57
Hematoxylin 116
Heptachlor 7, 13, 14, 16
— epoxide 3, 5
Heptane, purification 4, 7, 8
Herbicides 39 ff., 122
— and biochemical changes 39 ff.
— and enzymes 69—75
— and metabolic changes 39 ff.
— and minerals 58—63
— and photosynthesis 79—81
— and pigments 57 ff.
— and plant respiration 75—79
— chlorophenoxy 39 ff.
— concentration effects 84
— concentrations in cells 69 ff.
— fate in plants 83—91
— mode of action 92
— structure and activity 81
Hevea brasiliensis 56
Hexane, purification 4—8
HILL reaction 80
Hydrogen cyanide 21
— sulfide 21
Hydrolysis, automated 20

IAA effect on plants 39, 53—56, 64—
    70, 76, 85, 86
Illegal residues 17
Imidan 7
Infrared spectrometry 22
Intentional food additives 116—118
Interpretation of data 19
Ion-exchange, automated 20
Iso-octane, purification 7
Isopropyl alcohol, purification 6

Jimsonweed 87

Karathane 8
Kelthane, see Dicofol
Kidney beans 43, 44, 46, 48, 49, 53, 57,
    70, 72—74, 78, 88
Kinetin 53
Kuderna-Danish concentrators 3

Lambs quarters 46
Lamium anplexicaule 57
Larkspur 56
Lead 21
Legislative control of residues 14, 115 ff.
Lemna minor 45, 58
Lemons 28

Lepidium virginicum 57
Lethane 8
Light quality and 2,4-D effect 80
Limes 58
Lindane 16, 121
Linoleum 119
Linuron 16
Lipase activity 44, 45
Lipids 44
Lupines 56, 76
Lycopene 57

Macromolecules 119
Maize 52, 71
Malathion 4, 15, 16
Maneb 16
Market basket 17
Maximum acceptable daily intakes 122
MCPA effect on plants 39, 41, 42, 44, 45,
    47, 49, 54, 57, 58, 60, 63, 66, 72,
    80, 86, 87, 89, 92
MCPB effect on plants 40, 87
MCPP effect on plants 40, 89
Meat 120
Menthol 121
Mercury fungicides (see also specific
    names) 15, 16
Metabolic sinks 76, 92
Methoxychlor 3, 16
Methyl alcohol, purification 7
— bromide 15
— parathion 16
Methylene chloride 9
Methylenedioxyphenyl synergists 7
Microcoulometric gas chromatography 2,
    9
Milk and products 13, 120, 121
Millet 50
Minerals and herbicides 58—63
Minimum intervals 116, 121
Morning glory 40
Mustard 80

NAA effect on plants 39, 41, 53, 54,
    63, 85
NAD 40 ff.
$NADPH_2$ 40 ff.
Negligible residues 13
Netherlands legislation 15
— tolerance 31
Nicotine 16, 21, 55
Nitrogen metabolism 45—53, 72
— screening 18 ff.
Nitrophenylarsonic acids 15
Non-intentional food additives 118—123

No-residue registration 13
Nucleic acids 66—69

Oaks 44, 45
Oats 42, 44, 45, 47, 49, 50, 52, 63, 64,
    72, 75, 76, 80, 84, 90, 91
Olive oil 44, 72
Onethera biennis 63
Onions 48, 54, 65, 66
OPP 16, 117, 122
Oranges 28, 31, 32, 80
Organic acids 53-55
— farming 122
Organobromine compounds 18
Organochlorine compounds 20
Organophosphorous compounds 121
— AutoAnalysis 23—26
— residues, automated 23 ff.
Ornithogalum umbellatum 56
Orthophenylphenol, see OPP
Oxidation, automated 20
Ozone 70

Packaging materials 118-120
Paper chromatography 20, 120
Parathion 1, 5, 15, 16, 22, 23, 26
Parchment 119
Parthenocissus quinquefolia 63
Partitioning, automated 20
Peaches 23, 26, 57
Pears 43, 50, 55
Peas 41, 42, 50, 52, 53, 65, 67—70, 73,
    75, 76, 78, 85—87
Pectins 41, 53, 70
Pelergonium zonale 63
Pentachloronitrobenzene 7
Pentane, purification 4, 6, 7
Pentose-phosphate pathway 40 ff.
Peppers 41, 56
Permitted residues 13
Persian yellow 116
Persimmons 41, 57
Persistence curves 121
Perthane 8
Pesticide residues, symposium 118—125
Pesticides, registered 12
— types 15
Petroleum ether, purification 3, 5—8
Phenols 21
Phenoxyacetic acid 81
Phenoxybutyric acid 87
Pheophytin 57
Phosphates 21
Phosphorus screening 18 ff.
Photosynthesis and herbicides 79—81

Phytin 41
Pigments and herbicides 57
Pigweed 46
Pisum sativum 56, 60
PKDE, see Pyrocarbonic acid esters
Plasticizers in solvents 2, 8, 9
Polarography 22
Polyacrylate 119
Polyamide 119
Polyethylene 119
Polystyrene 119
Polyvinyl chloride 119
Potatoes 41, 45, 47—50, 52, 55, 57, 60,
    72, 73
Preservatives 115—117
— carcinogenicity 117
Processing 19
Protective colloids 119
Prunus spp. 51, 53, 75
Public health (Switzerland) 123
Purification, solvents 1 ff.
Puromycin 67
"p"-values, BEROZA 7
Pyrethrum 16
Pyrocarbonic acid esters 117

Quercus marilandica 88

Radishes 46
Raphanus sativus 49
Rats 78
Reduction, automated 20
Regulatory analyses 17
Residue analysis, automated 12 ff.
— analytical requirements 19—23, 120,
    121, 124
— regulation 12—15
— screening, automated 12 ff.
Respiration, plant 75—79
Rice 48, 63, 70, 72, 73
Ricinus 44
RNA 39 ff.
— synthesis 67
Ronnel 16
Rubber 4, 8, 9, 119
Rumex sp. 50

Sampling 19
Saponin 56
Scopolin and scopoletin 56
Silica gel 8
Silvex effect on plants 39, 44, 45, 54—57
Sinapis alba 43
Soapwort 56
Sodium in solvent purification 6

Sodium *o*-phenylphenol, see **SOPP**
Soil 18
— residues 13
— water 13
Solvents, chemical purification 2 ff.
— commercially available 2 ff.
— lacquer in 8
— purification 1 ff.
— soldering flux in 8
— specifications 5
— storing 2 ff.
SOPP 16, 117, 122
Sorbic acid 117
Sorghum 55, 87, 89, 90
Soybeans 46, 49, 51—54, 58, 61, 63, 66,
    67, 73, 78, 79
Spectrophotometers, automated 22
Split-blank procedures 22
Stabilizers 116—119
Starch 40—43
Steam distillation, automated 20
— biphenyl 28
Steroids 56
Strawberries 89
Strophanthidin 56
Sublimation 20
Sucrose 41, 42, 69
— and 2,4-D effect 75
Sugar beets 41, 44, 46, 49
— cane 41
Sulfur dioxide 21
— screening 18 ff.
Sulfurous acid 117
Sunflower 42, 43, 52, 56, 60, 63, 65, 73,
    74, 79, 80, 87
Sweet potatoes 62, 63, 72, 73
Swiss laws and regulations 123
Syringa vulgaris 89

2,4,5-T effect on plants 39, 41, 44—47,
    49, 50, 54—59, 62—66, 71—73, 82,
    84, 88—91

Taraxacum officinale 57
Tartar emetic 15
Tetradifon 16
Thin-layer chromatography 20, 120
Thiometon 16
Thiram 15, 16
Thistle 46, 47
Timothy 47
Tin 21
Tobacco 6, 55, 58, 66, 69, 70
Tolerances 12 ff., 116, 120, 121, 124
— international 15, 16
— violations 17
Tomatoes 41, 44, 45, 49, 54—60, 62,
    72, 73, 81, 84, 86
Toxaphene 4, 16
Triazines 15, 21
Triflorium hirtum 62
Triticum vulgare 60
Tropaeolum maius 63

Urea and ureas 15, 21

Vanillin 121
Vetch 77
Vicia faba 43, 47, 49
Vitamins 56
Vulcanized fiber 119

Water 13, 18, 63
Wheat 42, 43, 46, 49, 50, 52, 54, 58—
    60, 63, 70, 72—75, 86
Wine 117

Xanthophyll 57
Xanthosoma spp. 62

Zero tolerances 12—13
Zinc (see also Minerals) 21
Zineb 16
Ziram 16